CHEMORECEPTION IN MARINE ORGANISMS

CHEMORECEPTION IN MARINE ORGANISMS

Edited by

P. T. GRANT and A. M. MACKIE

Institute of Marine Biochemistry
(Natural Environment Research Council),
Aberdeen, Scotland

1974

ACADEMIC PRESS · LONDON · NEW YORK
A subsidiary of Harcourt Brace Jovanovich, Publishers

ACADEMIC PRESS INC. (LONDON) LTD.
24/28 Oval Road,
London NW1

United States Edition published by
ACADEMIC PRESS INC.
111 Fifth Avenue
New York, New York 10003

Library of Congress Catalog Card Number: 73 18999
ISBN: 0 12 295650 8

Text set in 11/12 pt. Monotype Modern, printed by letterpress, and bound in Great Britain at The Pitman Press, Bath

Contributors

Numbers in parentheses refer to the pages on which the author's contribution begins.

J. E. Bardach, Hawaii Institute of Marine Biology, The University of Hawaii, Honolulu, Hawaii, U.S.A. (49)

D. J. Crisp, Unit of Marine Invertebrate Biology, (Natural Environment Research Council), Marine Science Laboratories, Menai Bridge, Anglesey, U.K. (177).

P. T. Grant, Institute of Marine Biochemistry (Natural Environment Research Council), Aberdeen AB1 3RA, Scotland, U.K. (105).

M. S. Laverack, Gatty Marine Laboratory and Department of Natural History, University of St. Andrews, Fife, Scotland, U.K. (1)

H. M. Lenhoff, Department of Developmental and Cell Biology, University of California, Irvine, California 92664, U.S.A. (143).

K. June Lindstedt, Department of Biology, California Lutheran College, Thousand Oaks, California 91360, U.S.A. (143).

A. M. Mackie, Institute of Marine Biochemistry (Natural Environment Research Council), Aberdeen AB1 3RA, Scotland, U.K. (105).

Trudy Villars, Department of Psychology, The University of Michigan, Ann Arbor, Michigan, U.S.A. (49).

Contents

Introduction

The sense organs of animals function as highly selective filters for the enormous range of physical and chemical stimuli that can exist in the environment. Chemoreception means the sensory perception and discrimination of specific chemicals by animals and results in various behaviour patterns and traits. It is now accepted that chemical signals play an important role as mediators of inter-and intra- species relationships. An understanding of the fundamental reasons and principles underlying these behavioural patterns is essential for our comprehension of the complex parameters of population and other pressures that are concerned with natural selection and with the regulation of the dynamic equilibrium of a natural ecosystem.

The ancestry of most major animal groups can be traced back to early aquatic forms. Presumably, conditions of low intensity of light and a high degree of turbidity tended to favour and select the development of refined chemosensory mechanisms during the evolution of invertebrates. The first aquatic vertebrates to appear retained much of their dependence on the chemical senses and most marine animals have never developed the marked visual bias in sensory equipment that has evolved in man and some other higher vertebrates. In all animals, cells with a specialized chemoreceptor function, or chemosensory cells, have evolved where excitation and transduction can be evoked by specific chemicals. It is presumed that the transduction process results in an amplified output being transmitted from all chemosensory cells to the central nervous system where the impulses are modulated and analysed. In this way an animal possesses a discriminatory sense of awareness of changes in the chemical nature of its environment that complements and extends its other sensory armament. Most of our basic knowledge about chemosensory mechanisms has been derived from the study of a wide variety of stimulus-response systems. In most cases, the effects of specific chemicals has been determined with either a whole animal behavioural-response bioassay or with an isolated tissue preparation using the techniques of electrophysiology. The latter system has been extensively used to locate and determine the specificity of chemoreceptor cells and to identify the chemicals that *may* affect the behaviour of the intact animal. In this volume, the nature, specificities and functions of chemosensory cells in marine animals are discussed in Chapter 1 and with special reference to fishes in Chapter 2.

The release mechanism for instinctive animal behaviour patterns in response to chemicals is always identified with the selective absorption or filtration of the specific chemical by a chemosensory cell. At the same time, a specific chemical may trigger or "prime" a physiological process that ultimately results in an adaptive behavioural response. The concept that specific external chemical stimuli can function as "releasers" or "primers" of animal behaviour was initially developed by Wilson (1963) as one way of classifying intra-species communication between insects. These pheromones, or external hormones, were defined by Karlson and Lüscher (1959) as "secreted to the outside by an individual and received by a second individual *of the same species* in which they release a specific reaction, for example a definite behaviour or a development process". The isolation and structural identification of many volatile and odoriferous sex attractants and other pheromones of insects has been greatly facilitated by the microtechniques of gas chromatography coupled to a mass spectrometer. Some pheromones of marine animals have been identified but, in general, these water-soluble signals are present in only trace amounts and the isolation of sufficient pure pheromone from crude tissue extracts by *conventional* procedures is a difficult and daunting task.

Chemically mediated interactions between animals of different species have been termed allelochemic (Whittaker and Feeny, 1971) and are broadly concerned with chemicals secreted by one species that inadvertently or otherwise affect the growth, health or population biology of another species. A vast number and range of allelochemic effects have been detailed and described in ecological terms (chemical ecology) but a clear distinction must be made here between an effect such as the restriction of growth that results from metabolic inhibitors like antibiotics and toxins, and other allelochemic effects that are mediated by a chemical compound that functions as a releaser or primer on the behaviour of animals. It is the nature and role of these releaser and primer functions in marine animals that is the concern of this book.

Marine biologists and ecologists have provided valuable information about the range and scope of inter- and intra- species interactions that can be observed in the marine world. However, in most cases the involvement of the chemical senses could only be inferred. In recent years there has been an increasing emphasis on the study of these interactions under defined conditions and this volume consists of five reviews by different authors each reflecting, in part at least, their own research experience and authority. At the same time each review is concerned with the conceptual and experimental approaches that have been used to decipher the principles of chemoreception in marine animals. This is the first book to be wholly concerned with the subject and the Editors remain grateful to their distinguished colleagues both for the broad perspectives they have contributed and for their patience over the unavoidable delays that have occurred.

References

Karlson, P. and Lüscher, M. (1959). Pheromones; a new term for a class of biologically active substances. *Nature, Lond.* **183,** 55–56.

Whittaker, R. H. and Feeny, P. P. (1971). Allelochemics: chemical interactions between species. *Science, N.Y.* **171,** 757–769.

Wilson, E. O. (1963). Pheromones. *Sci. Amer.*, **208,** 100–114.

February, 1974

P. T. Grant
A. M. Mackie

I. The Structure and Function of Chemoreceptor Cells

M. S. LAVERACK

Gatty Marine Laboratory and Department of Natural History, University of St. Andrews, Fife, Scotland, U.K.

I. Introduction

It is not my intention in this article to provide an exhaustive review of all aspects of chemoreception and the properties of chemoreceptors. I have attempted instead to choose certain aspects in which recent developments have taken place and to use these topics to show general principles that may prove to be of value in providing a base from which future studies may extend. I have drawn my examples from numerous phyla and from varied habitats, believing that in the final analysis the necessity for chemoreception faced by all animals is basically the same, and has probably been solved in the same way. It so happens that insects, most of which are not aquatic organisms, provide especially favourable material for experimentation, and that consequently a great deal more is known about these animals than about some others. If therefore the species mentioned appear not to fall into the framework of chemoreception in aquatic organisms I hope this apologia will suffice as an explanation.

Nonetheless *all* chemoreception can be considered as essentially an aquatic phenomenon. In all recorded cases, including insects, there is some fluid material that surrounds the receptor ending. This fluid provides a buffer between the nerve ending (receptor cell) and the external environment which may be either aqueous or gaseous. It is probably true that all cells (including sensory cells) cannot survive in truly gaseous conditions owing to the extreme danger of desiccation by evaporation, and that some protection is necessary. Humidity detectors, where they exist, may be unusual in that some part of the cell membrane is directly exposed to the environment. In other types of chemoreceptor there is some material between membrane and exterior; this is invariably aqueous, and often in the form of a secretion.

Thus arthropod chemoreceptors are contained within cuticular hairs that protrude above the surface of the animal. The chemoreceptor dendrites end at some particular place within the shaft or in a pore at the apex. If they end within the hair they are immersed in fluid somewhat similar in composition to the external environment plus perhaps some extracellular fluid with proteinaceous substances; if the dendrites end at a pore, either this is intucked so that evaporation is much reduced, and presumably the relative humidity within the pore is high, or provision is made for the secretion of some fluid into the region. In other soft bodied animals the receptor surface may be totally immersed, and hence wetted, or mucous secretions from associated glandular areas may enable a viscous aqueous medium continuously to cover the receptor. Tongues, olfactory epithelia, carotid bodies, osmoreceptors, osphradia, nuchal organs and so on are all kept wet.

It is obvious that this provision of aqueous covering for the receptor ending is of great significance in any consideration of the mechanisms by which such an ending may function. The question of solubility of compounds in the fluid as well as the possible solubility in lipid at the cell membrane; the detection of water itself; and the presumably inert nature of water as a stimulant in most cases, are all topics that require examination in a full enquiry into chemoreception.

II. General Principles

The monitoring of chemicals within the environment sets a problem which appears to have been solved in similar ways by all groups of animals.

No matter what the immediate environment, be it blood or air, water or lymph, there must be some form of contact between the stimulant and the membrane of a cell (or the cell in the case of single-celled organisms). The form of energy provided as a chemical compound with all its attendant parameters of shape, size, solubility, Van der Waals forces, partition coefficients, vibrational energy and other physical chemical attributes is in some way active upon a portion (presumably a very small portion) of the living membrane that constitutes the receptive mechanism. This membrane is usually, on the Davson-Danielli model, assumed to be a lipid-protein bimolecular leaflet, though perhaps newer ideas (Sjostrand and Barajas, 1968) are challenging blind acceptance of this hypothesis (see also Stockenius and Engelman, 1969; Hendler, 1971).

This statement immediately indicates the first and greatest problem with regard to chemoreception, indeed in the field of chemical actions on cell surfaces in all aspects. How is the stimulating chemical detected and how are the consequent responses of cells related to this initial event? The act of reception by a chemically sensitive specialized organ may be defined as the transduction of one form of energy (physicochemical) into another (electrical). It is often thought to be a highly specific act in which a compound contacts a membrane, and by virtue of its shape, vibrational characteristics or size or chemical reactivity, in some way modifies the membrane so that the potential difference between the internal and external faces changes, usually in a graded manner.

Exactly how transduction takes place is not known for any animal or for any form of stimulation. The exact nature of a chemosensory membrane is not yet analysed, nor is there yet any knowledge of how such a membrane may differ structurally or biochemically from any other portion of membrane (but see later). In the past it seems to have tacitly been assumed that all membranes are the same, or if not the same at least similar. It can now safely be stated that they are not all the same, that although the biochemical structure may be broadly similar in all cells, it must be markedly different in detail. In nerve cells it is apparent that post-synaptic areas are responsive to the application of transmitters and other pharmacological agents whilst other regions are remarkably insensitive to such substances. It is likely that chemosensory cells may also have similar limited areas of sensitivity capable of interacting with chemicals in the environment. This may limit the area affected to a comparatively small fraction of the total cell surface although it is noticeable that chemically sensitive cells carried on specialized epithelial regions often have highly convoluted surfaces which greatly increase the total surface area.

The likelihood of different membrane structure, spatially arranged, means that there may be only limited regions with elaborately constructed molecules and molecular arrays that act as acceptor sites for materials impinging on them. In view of the considerable range of possible substances that may be detected in many systems (not by any means all since some are highly specific) it may be suggested that, rather than maintain a constant population of acceptor molecules at the surface, the cell continually replaces, destroys, modifies, repositions and resynthesizes the appropriate areas. This hypothesis would allow for a dynamic membrane that continually changes its properties within certain prescribed limits. Molecules presumably synthesized within the cell become incorporated into the membrane and are carried through the membranous layer to the exterior surface there to remain briefly to be destroyed by membrane-bound enzymes. There is no reason to suppose that compounds are placed in the sytem once and once only.

When in place at the surface the acceptor molecules act only as agents in the process of stimulation and response. The nature of the primary response as recorded physiologically is an electrical one, normally slow, graded, and decremental. If a substance impinges upon a cell membrane and is engulfed, linked, conjugated, chelated or combined with an acceptor molecule what happens then? There are few satisfying theories that enable us to conjecture how chemical energy is converted into an electrical phenomenon. Nearly all such ideas demand that some type of opening or gate appear in the membrane allowing efflux and influx of ions in differing concentrations and at differing speeds. Granted that the electrical event of the receptor is a reflection of ionic movement and that this necessitates aqueous solution on both sides of the membrane, how does membrane alteration take place?

III. Chemical Sensitivity

Human beings tend to underestimate the value and significance of the chemical sense to other animals. This is probably due to a comparative lack of sensitivity in human chemoreceptors, and the overriding importance of visual cues. Amongst the animal groups at large chemical sensitivity in many cases far outweighs in information content any other single physical attribute of the environment. It is often far more important than light, temperature, and mechanical agitation. This is as it should be for in very obvious ways organisms are dependent upon chemical information for their activities that cannot be mimicked by other cues. These can be considered generally as follows:

A. Feeding

The energetics of living systems, for the maintenance of bodily processes, synthesis of proteins, provision of energy, elaboration of secreted materials, and all other biochemical aspects of life, depend upon the provision of sufficient food in the correct proportions. Feeding is therefore at once the most important basic activity of all animals (even though some parasites do their best to overcome such obstacles by living totally immersed in food and hence have no immediate need for finding it; their problem comes earlier in their existence when they have to find a host in which to establish themselves). Feeding depends upon obtaining sufficient material from the environment for incorporation into the animal's own tissues and for metabolic energy. Since this demands the intake of other biological material (ignoring now some very special autotrophs) either plant or animal, it is of some adaptive consequence if the demand can be met with the least expenditure of existing energy supplies. Economy in such matters can be achieved if the food can be located and obtained with the least effort. Since all living things release a variety of compounds into the environment as the result of their own metabolic processes it is of great significance that such substances might provide clues for predators to home on to. Receptors concerned with responding to such compounds are commonplace in animals. Not all respond to the same compounds and some are highly specific, being concerned only in detecting one class or, or even just one, compound. Nonetheless it is the basic requirement of animals for food that has led to the evolution of highly organized chemoreceptors. As examples we may quote bivalves that increase their filtration rates when algae are in the ambient water (Jørgensen, 1955); many prosobranchs that search rapidly and accurately for their food when the conditions are correct (e.g. *Nassarius* moves to dead *Carcinus* (Crisp, 1971); *Buccinum* to *Mytilus*; *Urosalpinx* to *Ostrea*; *Aplysia* becomes active in the presence of *Ulva* (Frings and Frings, 1965); *Asterias* searches for mussels and barnacles (Castilla and Crisp, 1970)).

B. Escape

We are on less obvious, and less well known ground here, but it should be evident that whilst an animal needs food for itself and may take active steps to obtain it, particularly with chemical substances as a guide, there is also a necessity to ensure as far as possible that it does not itself become the food of some other organism. Escape responses are well described for a number of animals and whilst many are a result of visual stimulation probably a greater number are due to chemical stimulation. This may be the result of low concentration stimulation (as for *Asterias* in the same tank as *Solaster*, a predatory starfish), or relatively high concentration contact stimulation (e.g. *Pecten* fleeing from *Asterias*). Escape is an obvious adaptive response and serves to remove the presumptive prey from a predator (see Willows (1971) for a popular account of analysis in neuro-physiological terms). The extreme specificity of one such response is indicated by the work of Snyder and Snyder (1970) on *Diadema antillarum*. If a single *Diadema* is crushed or damaged, neighbouring urchins downstream move away as rapidly as possible. The response is species specific though some activity is noticed after crushing *Tripneustes* and *Lytechinus*. Natural predators such as *Cassis tuberosa* invoke fleeing behaviour only when chemical is combined with mechanical disturbance.

C. Reproduction

It has become more and more apparent in the last few years that aquatic organisms, in common with aerial and terrestrial dwellers, are in many cases reliant upon chemical cues in the environment for the processes of mating. In this case it is probable that very specific compounds are involved, possibly being quite restricted to a single species and a single sex. In aquatic animals that require active copulation (e.g. prosobranchs, decapod crustacea), and in those in which coordinated spawning is desirable (e.g. echinoderms, corals) the release of particular chemicals probably acts as a trigger to promote reproduction. A prerequisite, of course, is that the population in general be mature at the same time, and this is the concern of the internal chemical and hormonal environment, but the breeding behaviour and participation in mating is frequently released by chemical means. In cases where aggregations of animals occur at mating it seems probable that chemical lures are involved. It has yet to be determined in many animals how the signal is disseminated and over what distance it is effective.

D. Site Selection

Animals frequently favour particular areas if they are free ranging, whilst others are sedentary and hence unable to change their habitat for another region. It therefore seems mandatory in such cases that settlement and transformation into the adult should occur where conditions are most advantageous. This is particularly desirable at metamorphosis since the appropriate habitat for the larva is rarely suitable for the adult. The precise reasons for settlement of larvae in particular places has been determined so far for very few examples and in most cases biochemical agents responsible have not been isolated but it is clearly established that it is the chemical nature of the substrate that is overridingly important. The metamorphosis and settlement of *Ophelia* and of *Sabellaria* among polychaete annelids, and of the various barnacle genera amongst crustacea, have been shown to be dependent upon the very precise demands of the larva being met by the environment. In the case of *Sabellaria* this means that it must be in contact with the tubes of other adult *Sabellaria*, particularly the mucous component of the tube (Wilson, 1968); this predilection gives rise to the building of massive reefs as more and more larvae settle around the established adults.

E. Homing

A number of animals, especially molluscs (e.g. winkles, limpets etc.) are known to favour special sites on rocks for their "homes". These animals are grazers and browsers and forage for their food at a distance from the normal place of rest. They return, however, with regularity to the same spot. This ability is so accurate that eventually, for limpets at least, scars are worn on the rock and the shell fits exactly and precisely into this depression. The manner in which the animals relocate their original site suggests that a mucous trail containing a specific substance may be of value; possibly even specific to the single animal since the sites are personal and not interchangeable (Cook, A. *et al.*, 1969; Cook, S. B., 1969; Funke, 1968).

IV. Anatomy, Structure and Function

A. Situation of Receptors

The prime requisite of a chemoreceptor is that it must be in direct contact with that part of the environment that is to be monitored. This requirement distinguishes it from other modalities in which the receptor may be at some remove from the direct and immediate interaction of environment and sense organ, as in mechanically sensitive hairs, and eyes that are embedded deep in the body.

A demand of this kind is necessary because without its being met no chemical can interact with a receptor membrane. It also implies that in some way or other all cells are chemoreceptors for it seems probable that all are to a degree affected by alterations in the physical and chemical properties of their environment such as osmotic, ionic and hormonal changes. Thus hormonal responses of the mammalian liver, or the epidermis of young moulting arthropods are in a sense chemoreceptor responses. Equally all cells that line blood vessels, that are bathed in extracellular fluid and lymph, that are covered with fluid in an enclosed space (statocysts for example) may be susceptible to chemical change. Some of these are likely to be relatively slow changes with long time constants, and these we shall consider no further. We will restrict ourselves to those examples that are concerned with the rapid monitoring of chemical changes in the environment, which as a consequence of their immediate response give rise to propagated messages in neurones and which lead to some alteration in the behaviour (obvious or subtle) of the individual.

For our present purposes chemoreceptors may be defined as receptor cells specialized for the detection of small changes in the chemical composition of the environment due to substances volatile, soluble or electrolyte. Such cells are normally constituents of the nervous system, or are directly associated with the secondary neurones that act as conduction pathways to the CNS. They are often exquisitely sensitive. The cells need not necessarily be placed at the body surface, but may be incorporated in other organs, in blood vessels and in the central nervous system itself. They may also be effectors in their own right, as appears to be the case in the urophysis of teleost fish.

It is not surprising to find that chemoreceptors are to be found in almost any situation on the animal body. They may be single cells scattered over a body surface, they may be concentrated into special areas, they may be on hillocks exposed to the environment, they may be in pockets and cavities that are protected from the environment, they may be in tidal cavities (such as nasal pouches, or gill slits or blood vessels), they may be external or internal; indeed only scrupulously detailed examination of a species will reveal all the possible sites for chemoreceptor cells.

We know most about chemoreceptors which are gathered into aggregations, be they small as in insect labellar hairs, or large groups as in olfactory epithelia of vertebrates. Certainly the analysis of function is more readily undertaken in such areas where from behavioural studies it is shown that chemoreceptors must be present. In such cases electron microscopy may reveal the nervous supply, the origin of sensory axons and the nature of the receptor cell. Some examples are nevertheless not all that clear especially where several cell types occur within the organ and it is not possible to be specific about the function of each type when electron micrographs are the only evidence.

B. Histology and Fine Structure

Chemoreceptor cells are frequently small in size. In invertebrates they are normally primary sense cells and give rise to axons that reach the CNS. In vertebrates there may be accessory cells that synapse with afferent axons at their bases (see, for example, Whitear, 1971b).

Table 1 summarizes the information presently available on the fine structure of suspected chemoreceptors in aquatic animals. It is not completely exhaustive but other earlier references can be obtained upon resort to the papers quoted. One or two non-aquatic examples are included for comparative purposes.

It cannot yet be assumed that all the various types of chemoreceptor organs are described despite the many descriptions that are in the literature (see Bullock and Horridge, 1965). An example of how new additions may be made can be seen in the work of Szal (1971) in which re-examination of the ctenidium of *Tegula funebralis* (gastropod) revealed the presence of small swellings that are heavily innervated. Szal terms these "bursicles" and proposes a chemoreceptor function for them. They occur in Zygobranchia and Trochacea. The presumptive receptor cells are strongly ciliated, but further details are still awaited.

Chemoreceptor sense organs are found in all stages of the life history of animals. Larvae possess their own complement of such nerve cells and these must be as well adapted to the requirements of the juvenile stages as are those of the adult to the mature stage. Little attention yet seems to have been paid to them either structurally or physiologically except in some parasites, and in the barnacles. Some information is available on behavioural responses presumably mediated via chemoreceptors (e.g. Bayne, 1969).

Amongst parasites the chemosensory abilities of the larvae are of the utmost importance in finding a host, and thus it is not surprising to find arrangements of sense organs well disposed for such a purpose (Lyons, 1969; Chapman and Wilson 1970). Selection of a different kind, but equally essential, takes place in the metamorphosis of the barnacle larva from cypris to adult in *Balanus balanoides*. At this time the animal must obtain a suitable place to spend the rest of a sedentary life, having been nomadic until then. Nott and Foster (1969) have given a full description of presumptive chemoreceptors on the larval antennule with which exploratory movements are carried out. Only three ciliated hair endings are proposed as carrying chemoreceptors, but these have

Table 1. A summary of the information at present available regarding the structure of suspected chemoreceptors in aquatic animals (with one or two terrestrial examples for comparison).

Group: Phylum	Group: Class	Species	Organ	Dendrite Diameter μ	Cilia	Cilia length μ	No. cilia per cell	Basal body	Basal foot	Rootlets	Filaments	Filament arms	Other Structures	Axon diameter μ	Associated Cell Types	Other Possible Receptor Function	Author
Platyhelminthes	Turbellaria	*Dugesia tigrina*	Auricle	2	√		1	√		√	9+2				Epithelial cells	Mechano	MacRae, 1967
	Monogenea	*Gyrodactylus* n.sp (from *Gasterosteus*—Stickleback)	Terminal spike		√	11	several	√		√	14+5 variable				Mechanoreceptors collar	Mechano	Lyons, 1969
	Digenea	*Himasthla secunda* (from *Littorina*)	Cercaria papilla	2	√	3	1	√		√	? + 2					Mechano	Chapman and Wilson 1970
Nemertina	Heteronemertini	*Lineus ruber*	Cephalic	0·5—1·0	√		1	√		√	9 + 2				Neuroglandular vesicular ciliated lappet	Mechano photo Neurosecretion	Ling, 1969 (see also Ling, 1970)
Nematoda		*Haemonchus contortus*	3rd stage Juvenile Amphid		√	<18	1				10+(0—4)					Mechano	Ross, 1967
Annelida	Polychaeta	*Eurythoë complanata*	Nuchal		√		several	√		√	9 + 2				Support mucous	Photo mechano	Storch and Welsch, 1969a

		Nereis diversicolor	Prostomial Cirri: 1 Compound	7·5	√	7·5	7–15	√	√	9 + 2			Granular	Mechano	Dorsett and Hyde, 1969
			2 Solitary		No						Microvilli				
Arthropoda	Crustacea Copepoda	*Gladioferens pectinatus*	Mandible	1	No						Scolopale		Tormogen & Trichogen	Mechano	Ong, 1969
				2	√		1 ?	√	√	9 + 0	Scolopale			Mechano	
	Cirrepedia	*Balanus balanoides*	Cypris Antennule Hairs Axial		√		1	√	√	?	Scolopale		Supporting rods & tubes	Mechano	Nott and Foster, 1969
			Radial		√			√	√	?					
		Elminius modestus	Adult mantle		√								Hairs	Mechano	Foster and Nott, 1969
	Decapoda	*Coenobita compressus*	Antennular hairs	7	√	400	2	√	√	9 + 0	Branched cilia		Hypodermis		Ghiradella, Case and Cronshaw, 1968
		Pagurus hirsutiusculus	,,	7·5	√	1250	?	√	√	9+0(1)	Branched cilia		,,		Ghiradella, Cronshaw and Case, 1968
		Panulirus argus	Antennular hairs	0·2 – 0·5	√	7–8	1–3	√	√	variable 9+(1–4)		0·1 – 0·2			Laverack and Ardill, 1965
	Insecta Hemiptera	*Cenocorixa bifida*	Labial pegs		√	1·1	1	√	√	9 + 2 or more			Sheath	Mechano	Lo and Acton, 1969
	Arachnida	*Limulus*													Hayes, 1971

Table 1 (*contd.*)

Moll-usca	*Gastropoda* Proso-branchia	*Buccinum undatum*	Osph-rad-ium	1	1–2	No												
				2	1–2	√		sev-eral	√			9 + 2		Micro-villi	0·1—0·5		Mechano	Welsch and Storch, 1969
		Neptunea antiqua		3	1–2	No												
		Nassarius reticulatus	Siphon & ten-tacles		1·5	√		2–10	√		√	9 + 2		Micro-villi	0·1—0·5		Mechano	Crisp, 1971
	Nudi-branchia	*Polycera quadrilineata*	Rhino phores		0·5—1·0	√	sev-eral											Storch and Welsch, 1969b
		Adalaria proxima																
		Facelina drummondi																
		Dendronotus arborescens																
	Cephalo-poda	*Nautilus macromphalus*	Rhino-phores flask cells		20	√	1 or 2 are long	many	√	√		9 + 2		Micro-villi		Ciliated Mucous		Barber and Wright, 1969
		Octopus vulgaris	Suckers		1–3	√	2–3	1–5	√	1	√	9 + 2	√	Micro-villi				Graziadei, 1964
Verte-brates	*Cyclo-stomes*	*Lampetra fluviatilis*	Olfac-tory		5–6	√	5–6	6–10	√	>1		9 + 2	√		0·15 0·5	Basal Support abnormal		Thornhill, 1967
	Teleost	*Carassius carassius*	Olfac-tory		8	√	5–15					9 + 2		Com-pound cilia	0·5 0·7	Support mucous basal		Wilson and Wester-man, 1967
		Lota lota	Olfac-tory		2	√	20–30	c20	√		√	9 + 2			0·15	Basal glia support		Gemne and Døving, 1969
		Phoxinus	Olfac-ory	1	1·5	√	5	4–6	√		No	9 + 2	No			Support ciliated		Bannister, 1965
				2										Micro-villi				
				3										Rod shaped				

	Gasterosteus aculeatus	Olfactory	1·5	√	5	4–6	√		No 9 + 2	No			Epithelium	Bannister, 1965
	Clarias batrachus	Taste buds	1–2 2 types	No							Microvilli		Synapse onto nerve	Welsch and Storch, 1969
	Kryptopterus bicirrhis													
	Phoxinus phoxinus, *Gasterosteus aculeatus*	Taste buds		No							Microvilli			Whitear, 1965, 1971 a, b.
	Trigla lucerna, *Pomatoschistus minutus*	Epidermis												
Amphiba	*Rana clamitans*	Olfactory		√	up to 200	4–8	√	√	No 9 + 2 (65%) 8 + 2 (16%) and others	√			Support	Reese, 1965
Reptilia	*Anguis fragilis*	Vomeronasal		No							Microvilli			Bannister, 1968
	Gopherus polyphemus	Olfactory		√	10–15								Support	Graziadei and Tucker 1970
	Terrapene carolina	Vomeronasal	1·5	No							Microvilli	0·14–0·4		Graziadei and Tucker 1970
	Lacerta	Vomeronasal		No							Microvilli	0·17		Altner and Muller, 1968
Aves	*Anser*	Olfactory	2–3·5	√	12	8–10	√	>1	No 9 + 2		Microvilli	0·1 0·25	Support Basal	Graziadei and Bannister, 1967

pores at their ends and are located at strategic places. No remarks as to their sensitivity can yet be made, but it seems likely that they are concerned with detecting the presence of arthropodin (Crisp and Meadows, 1962) which guides barnacle larvae in settlement.

The adult barnacle also possesses hairs that contain ciliated dendrites within. These hairs are found at the mantle edge and are suitably placed for monitoring the environment beyond the edge of the shell. It is suspected that these may be chemoreceptors, but the possibility also exists for a mechanical sensitivity (Foster and Nott, 1969).

Other arthropods upon which some detail is available include several decapods, including crayfish (*Austropotamobius pallipes*), crabs and lobsters. Thomas (1970) gave a full account of all the setae (hairs) of the crayfish, an exhaustive description of various types of outgrowths found on all the various limbs and appendages adapted for various functions. In a number of these, including the aesthetascs of the antennules there is a pore at or near the end of the hair. Thomas does not comment upon the innervation of these hairs so it is not possible to do more than speculate about the inclusion of chemoreceptors in any, though the hairs seem especially well represented at the mouthparts and setal types associated with feeding seem to have pores.

Although arthropods carry the chemosensory dendrites in hairs or other projections of the body it is still apparent that the external environment must in some way affect the dendritic membrane.

Volatile materials impinging upon the external cuticle must either contact the cell membrane directly at the surface of the animal or penetrate the cuticle to affect the membrane contained internally. Initial studies with the electron microscope using insect examples seemed to indicate that the dendrite extended as fine branches through small pores in the cuticular wall to end in bulbous cavities in communication with the interior. More recent studies show that the pores are in reality intuckings of the thin cuticle than push into the internal follicular or luminal fluid that fills the hair. This allows two possible sites for interaction or stimulating molecules with the receptor cell. Either the tubules make direct contact with the dendrite and molecules are brought immediately to the cell surface, or alternatively the stimulant substances leave the pore tubules to dissolve in the fluid of the hair canal and thence reach the receptor cell (Steinbrecht 1969b).

In this debate comparative evidence may be obtained from the crustacea. Laverack and Ardill (1965) showed that the distal region of the aesthetasc hair on the antennules of *Panulirus argus* contained no cellular extensions but was apparently filled with fluid that probably contained at least some protein which precipitated upon fixation. These authors believed that a pore exists at the distal end of the hair allowing ingress of the external watery medium into the lumen of the hair. On this assumption the stimulant chemicals would need to pass down about 60% of the hair length (several hundred μ) to reach the receptor dendrite. Studies with a scanning electron microscope of the aesthetasc would confirm or deny the presence of a pore in these hairs.

Ghiradella, Case and Cronshaw (1968) in their studies on the hermit crab, *Coenobita*, and other decapods, found no pores in the structure of the hair. Nonetheless cresyl violet and other dyes pass across the cuticle into the hair. If appropriate stimulatory chemicals can also pass through the cuticle the route must be via the luminal fluid to the receptor. Lo and Acton (1969) find the same organization in the water bug *Cenocorixa bifida*.

In arthropods therefore it seems likely that in both aerial and aquatic forms the essential mechanism of stimulation is by transference of stimulants from the exterior across the cuticle to the internal fluid and thence to the receptor-dendrite. In both cases the chemical is presented in the fluid phase to the dendrite. It seems likely that a general principle exists in the necessity for all chemical stimulation to occur as an aqueous solution even in the highly terrestrial insects.

In decapod crustacea those hairs that have been definitively identified as in some way concerned with chemoreception all seem to be innervated by enormous numbers of dendrites. Laverack and Ardill (1965), Ghiradella, Case and Cronshaw (1968) and Ghiradella, Cronshaw and Case (1968) all report that for aesthetasc hairs on antennules 300–400 dendrites per hair is a usual number. Shelton and Laverack (1970) indicate similar numbers in the stout hairs of the first and second walking leg chela in *Homarus gammarus*. Ghiradella, Cronshaw and Case (1968) believe that all dendrites receive information from all parts of the hair (because they branch) and that this constitutes a form of peripheral convergence. No synapses or nexuses have been described but tight packing and large numbers may allow electronic spread between units. In this case each separate nerve cell would interact with many others and hence spread information rather than channel it (cf. Steinbrecht 1969a).

In the very much smaller copepod *Gladioferens*, Ong (1969) reports two types of receptor, but unfortunately is unable to characterize either of them as chemically sensitive on physiological grounds. One of these organs carries a cilium, the other does not; the non-ciliary type ending at the surface in a slightly bulbous structure. One has a centrally placed cell soma. In situations like this it is not easy to be certain about cell function.

The same problem arises in annelids (polychaetes) in which Dorsett and Hyde (1969) have reported two distinct receptors on the prostomial cirri. Either or both of these may be chemoreceptors. They are particularly interesting for being of very different types and, if both are chemoreceptors, this parallels the situation in vertebrates. One is ciliated, the other microvillous in structure. In vertebrates this distinction seems to mark off olfactory or low threshold organs from taste organs or contact receptors. If this is so it argues for different cellular origins for these two functions.

Several cell types have been intimated by Welsch and Storch in a series of papers (see Table 1), concerned with molluscan, annelidan and vertebrate sense organs. In most cases there is insufficient detail to be sure of the identification of nerve cells and others, and there is no physiological evidence to substantiate the claims made for function of the cells. Crisp (1971) provides more satisfactory results for *Nassarius*.

Vertebrates demonstrate two receptor types as mentioned above, one ciliated the other not. These two types may correspond to olfactory (low threshold) and contact (high threshold) receptors since the distribution and occurrence is as expected for such a proposition.

Consideration of Table 1 invites several general comments.

1. The origin of chemoreceptors may be polyphyletic. It is apparent that the cilium has been an important evolutionary precursor for a basic receptor structure. In the case of mechanoreceptors, photoreceptors and chemoreceptors the basic ciliary structure has been utilized many times. It can now be safely postulated that the cilium of the protists has evolved into a variety of sensory types. From an originally motile organelle it may have taken up a number of forms, with a variable internal structure, but all subserving some sensory function. In the case of mechanoreception Thurm (1968) has shown that motile cilia retain a certain mechanical sensitivity and respond to lifting by another rapid beat cycle. Why the cilium should represent such versatile membranous material is unknown.

The cilium represents one type of chemoreceptor ending that carries transduction sites. The second type has arisen from a different cellular origin, or from a greater reliance on the general unspecialized cell surface. Particularly amongst the vertebrates, as in taste buds of fish and other groups, the sensory cell is non-ciliated but bears an array of microvilli. Some presumptive chemoreceptors amongst invertebrates also carry microvilli. Cilia and microvilli both increase the cell surface and the available sensory area.

2. In cases in which the cilium seems to be involved in chemoreception the number of internal filaments found is variable. The basic $9 + 2$ arrangement seems common and resembles the motile cilium closely although the subfibril arms may not be regularly present. This is a feature, however, that has been overlooked in many reports. In other

sensory cilia a 9 + 0 array is common, but is apparently unusual amongst chemoreceptors.

3. It is frequently difficult, especially in the case of scattered receptors, to establish the truly chemical sensitivity of cells. In organs consisting of collections of sensory cells it may be relatively simple to demonstrate chemosensitivity, hence the structures of receptor cells may be equated with function easily; but in other cases function is not so readily demonstrated and thus possible alternative receptor functions must be examined.

4. In all reported cases the diameter of chemosensory axons is very small. There are thus large numbers of axons to be found in comparatively slender nerves. The functional reasons and consequences are obscure but compound potentials due to mass firing and synergism due to cross talk seem quite possible.

5. The basal foot, protruding to one side of the ciliary basal body only appears in advanced forms (*Octopus*, cyclostomes, birds) and in these cases may be multiple (although there are undoubtedly exceptions to this rule, e.g. *Nassarius*). The significance of this can only be conjectured though obviously the directionality of motile cilia, sometimes thought to be consequence of the basal foot arrangement, cannot be a prime factor here.

6. Microvillous receptors versus ciliary receptors *may* be indicative of distinctive threshold levels. The table shows that both types may be found in the same animal, even in the same region, but often in different organs. Thus in vertebrates olfactory organs carry ciliary endings, and subserve distance chemoreception, whilst taste buds (essentially contact receptors) are microvillous. In *Nereis* prostomial cirri both types occur. These may be physiologically distinct and concerned in perception either at a distance or in palpation of food close at hand.

7. There are often enormous numbers of chemosensory units situated close together and acting in parallel. But where there are few there seems to be no diminution of discriminatory powers, as in insect hair sensilla.

8. Where on the cell the initial steps in the olfactory response occur is not clear. Generally it must be suspected that the membrane surface is important, and yet Tucker (1967) reported that 99% of the olfactory cilia in box turtles may be removed without abolishing the response of the organ to amyl acetate. The basal bodies of the cilia were still retained. If cilia are of functional importance rather than being an evolutionary relic then the presence of their peripheral regions is not obligatorily essential for receptor function. If it is postulated that acceptor sites are carried on the membrane and that the membrane is increased in area by intuckings and outgrowths then those that are left

after removal of all excrescences are still sufficient to promote the response (see page 41).

9. In *Phoxinus* (teleost) a third type of receptor has been described (Bannister 1965). It remains to be seen if this represents yet another form of chemosensory ending.

C. Replacement of Sensory Cells

In aquatic vertebrates the sensory cells are periodically replaced. In *Lampetra*, degeneration of the cells and differentiation and regeneration of the ciliated endings has been reported by Thornhill (1967). The recent classic studies by Young (1971) on the continual growth and replacement of the outer segment of rods in the vertebrate retina using high resolution autoradiographic methods would be a useful confirmatory technique for these observations.

There is no suggestion at present that similar replacement occurs in invertebrates though it seems a distinct possibility that regeneration of sensory endings must take place in some arthropods after moulting. Although the moulted cuticle has already been replaced before ecdysis occurs some attrition of fine sensory endings may take place. The value of a centriolar apparatus in remoulding distal sensory endings may be an evolutionary reason for retention of cilia in this capacity even in phyla that otherwise lack cilia. In the majority of invertebrate receptors, however, it is assumed that the endings are permanent, although the likelihood exists that all surface protuberances may be worn or torn away.

D. Chemoreceptor Nerves

The axons of chemoreceptors, both olfactory and contact types, seem to be uniformly of small diameter and unsheathed. Bundles of axons may be collected together in sheaths, but individual fibres are not separated from their fellows by any interpolated cellular structures. The average diameter of such axons commonly lies between 0·1 and 0·5μ (Fig. 1). Electrical stimulation of the whole nerve is normally followed by the appearance of compound action potentials, being the summed potential of many units responding to the stimulus. These are slow (C fibres in vertebrates) with a conduction velocity of 0·1–0·2m/sec. Slow conduction is common to other groups of very small fibres also, such as the echinoderm radial nerve cord (Sandeman, 1965).

Easton (1965) working on the olfactory tract of the garfish (*Lepisosteus osseus*, *L.platyrhynchus* and *L.oculatus*), reports that for such fibres the impulse length (length of nerve involved in the generation and propagation of the action potential) is very small, of the order of millimetres, and the space constant is small but the impulse duration

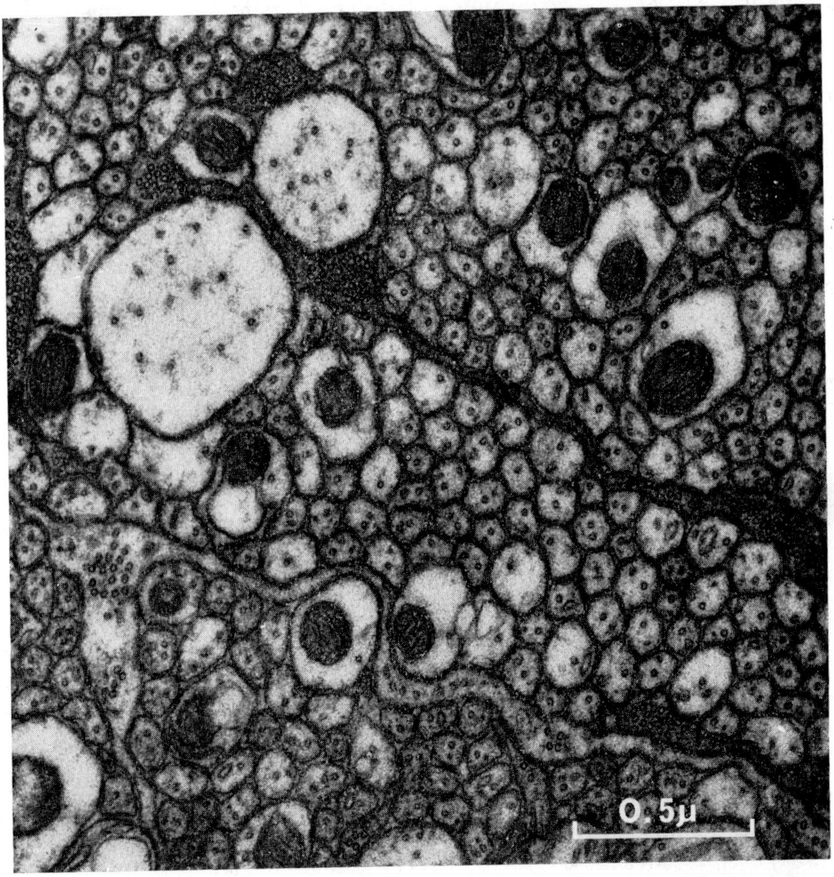

Fig. 1 Section through the antennal nerve of *Rhodnius prolixus*. There are large axons with no separate glial sheath, and many small diameter axons. Glial processes permeate between bundles of axons. This is typical of a section of a chemosensory nerve and is indicative of the very small size of the majority of axons. (By courtesy of R. Steinbrecht, to whom I am grateful.)

(time) is very long. He also reports that slow potential changes accompanying action potentials may be long lasting, enduring for up to 30 secs after a short period of tetanus (10Hz for 10 sec).

E. Synergism and Axon Fusion

The electrophysiological recording of potentials from a chemoreceptor population gives two distinct and separate sets of information: one on the activity of single units within the population, and one on the change in potential of the group as a whole. These may well be opposed with single units actively discharging whilst the DC potential shows a hyperpolarizing (or inhibitory) phase.

Synergism of action, i.e. the response of groups of axons simultaneously in the same way, might be explained in some cases by synchrony of firing or by the massive interaction of ionic movements in the group. Early work suggested that in some animals, notably arthropods and annelids, axons originating at peripheral receptors may be separate at their origins, but at some more central position fuse, and hence give complex possibilities for integration at the periphery. Synergistic responses could be a result of such fusion. It is therefore relevant to note that Horridge (1963) has suggested that in the polychaete annelid *Harmothoë* the segmental nerves seem to carry a one to one axon to receptor projection of any sort (a view disputed by Lawry, 1968, 1970). Better evidence for maintained integrity of sensory axons comes from Steinbrecht (1969a) in an analysis of numbers of nerve fibres in *Bombyx* and *Rhodnius*. His estimate of cell numbers coincided with numbers of sensory axons, and hence he concluded that these axons do not fuse; they do, however, run in bundles of unsheathed axons close together, and hence there exists the possibility of cross talk. Steinbrecht points out that in *Bombyx* there are 30,000 sex attractant receptor cells each connected separately to the deutocerebrum. Stimulation of these leads to great behavioural consequences, even when a very small stimulus, possibly stimulating very few neurones, is delivered. Under such circumstances cross talk may be of significance.

The question of information flow and number of channels in these insects seems to be of a different order of magnitude from that in vertebrates, for example in the fish *Lota lota* where Gemne and Døving (1969) estimated that 11,000,000 receptors each give rise to an axon 0·125–0·150μ in diameter. There is here, however, a convergence factor of 1000:1 of olfactory fibres onto secondary neurones with myelinated axons. Efferent fibres also exist in the olfactory nerve. Redundancy and interaction between channels may be considerable.

V. Chemoreceptor Responses

Electrophysiological recording is a rather hit and miss method of detecting chemoreceptor activity. As we have seen above the numbers of neurones involved may be astronomical. It is therefore a chancy business to record from identifiable units except in favourable circumstances. This is the major reason for the privileged position of insects in experimental analysis.

The osphradium of molluscs is a suspected chemoreceptor that because of its morphology and anatomy is peculiarly difficult to investigate by direct means. Electrophysiological experiments designed to record activity have given sketchy results (Bailey and Laverack, 1966). It has, however, been possible to show effects of stimulation of the peripheral organ upon central neurones. Bailey and Laverack working on *Buccinum* showed excitation of central neurones after stimulation of the osphradium with extracts of *Mytilus*. Jahan-Parwar *et al.* (1969) found that the "parabolic burster" unit of Strumwasser (R15 of Frazier

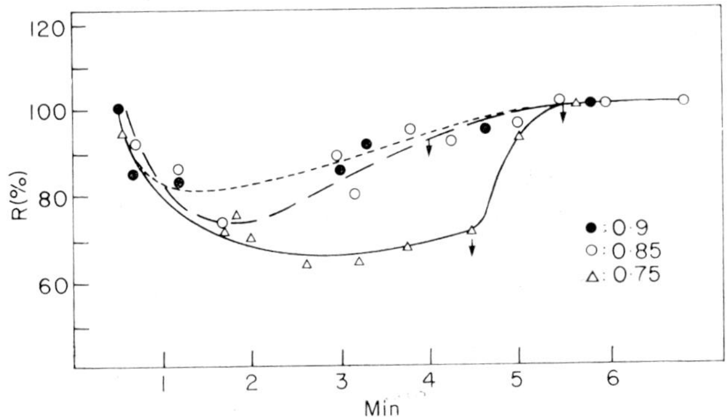

Fig. 2 The cell R15 of the abdominal ganglion of *Aplysia* is normally active in a rhythmic bursting fashion, generating small trains of spikes at more or less regular intervals. At certain periods, however, the unit may be silent. At such times application of dilute sea water to the osphradium affects R15 by a direct pathway which leads to a decrease in the membrane resistance of the cell. In the case shown relative changes in membrane resistance are plotted as a function of time during immersion of the osphradium in dilute sea water. Arrows show the time of return to normal concentration. Similar experiments demonstrate a variation in membrane potential; and during the rhythmic firing period the discharge is inhibited. (From Stinnakre and Tauc, 1969.) (Reprinted with permission of the Journal of Experimental Biology.)

et al. 1967) in *Aplysia* may be turned off (inhibited) by various peripheral events such as osmotic changes, extracts of seaweeds and shrimp pellets. Other neurosecretory cells like R15 may be excited or inhibited according to their active state. Stinnakre and Tauc (1969) confirmed and extended this observation on *Aplysia*, especially in response to hypotonic sea water (Fig. 2). Clues as to the nature and response characteristics of receptors may therefore be available from distant sources.

Those interested in chemoreception amongst marine invertebrates, and wishing to see the range of information available should refer to Laverack (1968). Attempts to record directly from chemoreceptors in aquatic invertebrates have been rather few; the main information coming from the larger crustacea.

Intracellular recording is by no means easy in chemoreceptors. This is a difficulty imposed by the extremely small size of the individual cells of the organs. In most cases the reported diameters of the somata are between 5 and 15μ whilst the axon (and dendrite) diameters may be very much smaller than this (down to $0{\cdot}1\mu$).

Altner and Boeckh (1967) used refined techniques to record from the olfactory epithelium of frogs and were successful in obtaining responses from individual sense cells, recording spike potential discharges. They did not however obtain intracellular penetrations. Sato (1969), however, has reported successful penetrations of taste cells in frogs (*Rana nigromaculata* and *R.catesbiana*) in which the chemosensory cells are 7μ in diameter. By recording intracellularly in the appropriate cell layer of the tongue, Sato believes he has recorded receptor potentials in response to chemical stimulation. Other cells in other situations showed no response to chemicals applied to the tongue. The cells that Sato impaled, showed a resting potential of 10–35 mV and penetrations were held for a period ranging from a few seconds up to 10 mins. In all, the receptor potential appeared as slow depolarizing potentials, never exhibiting propagated spikes. Using sodium chloride, acetic acid, sucrose and quinine as stimulants Sato demonstrated that each cell has multiple sensitivity and is not specific to any one type of substance.

In view of the previous description and suggestion regarding types of sensory cells it is possible to further speculate that where there are two distinct dendritic endings there are also two distinct primary responses. The ciliated receptors typical of the frog olfactory epithelium actively propagates potentials; the microvillous receptors of the tongue do not.

We must assume in the absence of any clearly contrary information that chemoreceptor cells function in the same or similar manner to other excitable cells. That is, upon stimulation by some appropriate compound, exchange of ions occurs across the cell membrane. These ions no doubt fall in the common categories of Na^+, K^+, Ca^{2+}, Mg^{2+} and Cl'. Studies on certain mechanoreceptors show that Na^+/K^+ exchange occurs and that this is comparable to the events noted for axons and nerve cells.

Problems of analysis of ionic movements arise in most chemoreceptors, possibly in all, and there is as yet no clear explanation of what ionic events take place. Experiments on frog olfactory epithelium (Gesteland *et al.*, 1965) and on the very different preparation of an insect chemoreceptor (Boeckh, 1967) have shown that both excitation and inhibition (or depolarization and hyperpolarization) may occur at the same time in the same preparation. In the frog the two types of slow potential recorded from the whole epithelial layer may be of opposite sign with one process seeking to return the potential to a certain level or even positivity, whilst the other change displaces the potential in the negative direction. This may be correlated at the same time with changes in spiking activity of the individual receptors. Boeckh (1967) and Wolbarsht and Hanson (1967) demonstrated that during recording from any one cell adjacent cells may have different response characteristics. Stimulation by excitation of a specific chemoreceptor may be paralleled by a simultaneous hyperpolarization of another receptor; so the generalized receptor potential reflects the changes in polarization

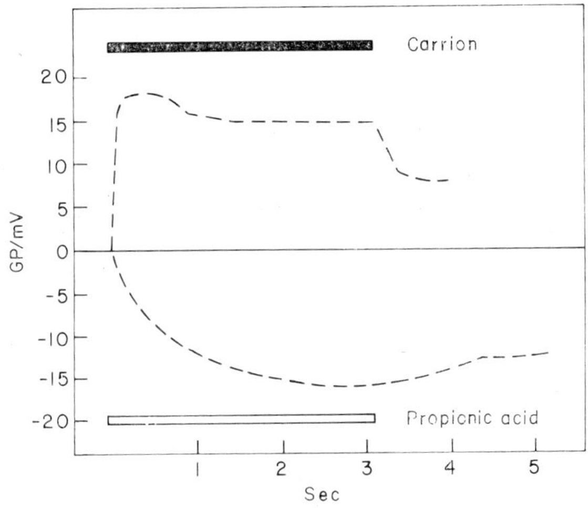

Fig. 3 Receptor potentials recorded from the carrion receptor of *Thanatophilus rugosus*. Carrion (shown by black bar) leads to depolarization of the receptor cell; propionic acid gives rise to hyperpolarization (inhibition). The 0 line indicates extracellular potential of unstimulated preparation. (From Boeckh, 1967.) (Reprinted with permission of Pergamon Press Ltd.)

across all receptor membranes which may be different from that of any one particular unit. Boeckh (1967) shows that different stimulating materials evoke different responses that may both excite or inhibit by changing the membrane potential of a cell (Fig. 3). The two potential changes have different time courses and hence the same cell may both fire action potentials and also hyperpolarize at different times.

Synaptic events have also been shown by Hughes and Tauc (1965) (in *Aplysia*) to demonstrate excitatory and hyperpolarizing phases (BPSP) which may be due to the activity of two transmitters at the same site, invoking differential ionic movements.

In chemoreceptors then the action of the stimulant upon the receptor membrane brings about membrane conductance changes that we assume are equivalent to those of axons. We have no definite evidence so far that this is so. It may, however, explain the necessity for aqueous substances to be available at the receptor dendrite region (even in insects) since ionic availability presumably depends upon aqueous solution being present. The composition of such a solution may be dependent upon the nature of the blood and body fluids, and in some cases the aqueous environment only. Mucous secretions are well known as overlying receptor endings; this may serve a double function as first a solvent for the stimulating substances, and second as a source for ions. The actual composition of the mucus may be very variable, but presumably this is not critical.

The complex impulse patterns recordable from chemosensory organs carry information to the CNS but there are few cases of detailed analysis of the value of this information to the animal, and how it is utilized. That there may be modification we already have seen, but Dethier (1968) shows that this may be of great significance in the normal life of the animal. The response of the animal (a fly) depends upon its previous history. After the fly has fed to satiety upon water it no longer shows proboscis extension to water or salt. A fly previously unfed responds to water or sugar by extending the proboscis and feeding. High concentrations of salt lead to failure of this reflex and withdrawal of an extended proboscis.

The reflex is very rapid and acceptance or rejection is almost immediate (no rolling of wine on the tongue, or savouring the aroma of a dish before partaking). A difference in receptor response of only 3 impulses in the first 100 msec leads to either of the behavioural responses. Dethier postulates a rate detector in the CNS responding to changes of about 20% in frequency. The rate detector activates either one or two neurones; these have different thresholds so that response to the lower input results in acceptance, to the higher leads to rejection.

The precise nature of a receptor response is therefore of great importance, but simply looking at filmed records does not tell one what the significance to the animal may be.

VI. Central Nerve Control

Possible direct overriding control of chemoreceptors exists in many active predatory animals. That is to say feeding is an action that occurs periodically and once accomplished need not be repeated for some period. If *Homarus gammarus* is fed it will not feed again until hungry although the food is available *ad libitum* (Laverack and Wales, unpublished observations). The same is true of divers organisms such as

Acanthaster planci (Asteroidea; Barnes *et al.*, 1970), *Nassarius reticulatus* (Prosobranchia; Crisp, personal communication) and sundry cephalopods.

Several possibilities exist for control mechanisms. Efferent control of receptors from a central source has been well demonstrated in one preparation from a truly aquatic animal, that of the olfactory bulb of the burbot *Lota lota* (Døving, 1966; Døving and Gemne, 1966). Electrophysiological recordings made from the central end of a cut olfactory tract showed impulses travelling towards the periphery, in a spontaneous manner. This activity was increased by stimulation of the ipsi- and contralateral tracts. The nerve cells involved probably have their cell bodies in the telencephalon, and possess myelinated fibres. The sensory cells of the bulb may be either excited or inhibited (60%) by the information in efferent fibres; these responses were followed by a period of poststimulatory inhibition which might be prolonged in duration. The long latent periods recorded in some cases suggest that bulbar integration of some sort occurs, and that efferent activity is carried in high order neurones.

Control of behaviour by feedback may be generated by a full stomach (monitored by proprioceptors) that passes either directly, or via some intermediate pathway in the CNS, back to the receptor system (see Dethier and Gelperin, 1967; Gelperin, 1966). Yates *et al.* (1970) have shown afferent neurones ending at the carotid body in close proximity to the glomus (Type 2) cell which seems to be the chemoreceptor in this organ. Stimulation of the sinus nerve leads to release of amines from these cells though there is as yet no indication of the function of such release.

Esakov (1961) working on frogs demonstrated that receptor output from the tongue was modified by alteration of internal conditions. Thus distension of the stomach by water led to an increase in the afferent discharge of water and salt stimulated units. If, however, a solution of peptone in water was introduced into the stomach responses to water and salt solutions on the tongue were diminished. In both cases quinine effects were largely unaltered. The reflex nature of these responses was indicated by their abolition following application of cocaine to the stomach and by extirpation of the medulla.

It seems a well founded observation therefore that feedback occurs which modulates the activity of the receptors, mainly by inhibition levels which are regulated from the CNS. This provides a method of fine tuning or biassing of receptors. How this is achieved at the receptor is not yet known (Fig. 4), but nerve endings at the base of sensory cells have been described (e.g. Whitear, 1971b), and synaptic interaction may be expected.

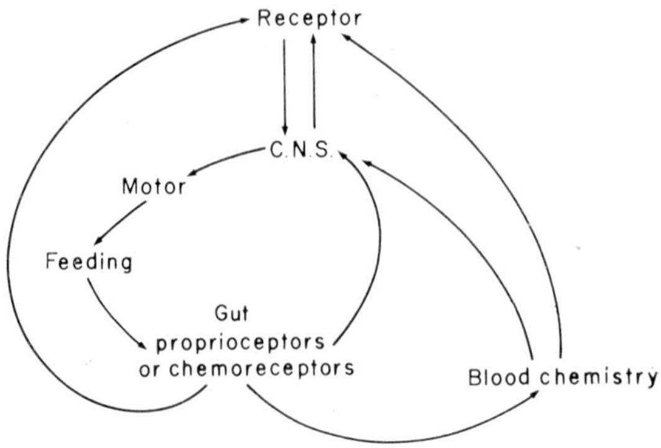

Fig. 4 Diagram to show possible pathways of feedback control of peripheral receptors. Stimulation at the periphery leads to changes in the central nervous system, which is followed by motor output and feeding. Intake of food leads to changes in volume of the gut and the content may also be monitored by internal chemoreceptors. These receptors may then affect the externally placed receptors either directly or indirectly via the CNS. Alternatively changes in blood chemistry may alter conditions around the CNS or receptor.

VII. Stimulatory Compounds

A. Individual Substances

It is of course impossible to do more than generalize about the chemical substances that may stimulate any chemoreceptor. There may be extreme specificity as in the case of pheromone receptors, or great lack of specificity in other cases. The theories of sterochemistry for vertebrates bring compounds into classes but there are numerous exceptions to the generalization (Beets, 1970). No one has yet attempted to classify stimulants in invertebrates in quite the same way.

Numerous different observations have been made on sensitivity of animals towards chemicals, especially by behavioural methods. I shall restrict myself here to a few indications of the recent approaches that have been made.

Leeches feed by attaching themselves to the host, incising the body wall and sucking blood. Galun and Kindler (1966) showed that leeches will attach themselves to a membrane of ox caecum and will feed on the contents of whole blood, washed erythrocytes, plasma etc. Maximal weight gains (feeding to satiety) were obtained with plasma (761%) blood (627%) and washed erythrocytes (685%). Less weight was gained by leeches fed with salt and sugar solutions and NaCl alone gave an increase of only 66% on the starting weight; glucose gave no change. The garument in this case is for combination effects on receptors, with high molecular weight organic compounds in the blood accounting for a considerable fraction of the uptake. Sugar specific receptors may, however, be present since phlorizin inhibits the sucking response. These responses do not demonstrate how the leech is attracted in the first place. This might be by a combination of mechanical and chemical stimulation.

Attraction of specific type may be a very commonplace phenomenon in feeding of animals. McBride, *et al.* (1962) showed that sockeye salmon (*Oncorhynchus nerka*) can be conditioned to beef extract and to zooplankton. Aqueous extracts of these materials injected into the experimental tanks increased the swimming rates, exploratory and feeding responses, thus indicating a sensitivity towards the extracts. Extreme sensitivity is shown in the same species by responses to "home water" taken from the stream in which the fish was bred. This is probably an imprinted response derived from very early conditioning. Work on the "homing" of salmon has been reviewed by Hara (1970). Conditioning has also been successful in the gastropod *Urosalpinx* (Wood, 1968).

In fish indirect evidence as to the possible sensitivity of receptors to chemicals has been given by Hodgson *et al.* (1967) who studied the electroencephalograms of free swimming sharks (*Negaprion brevirostris*, lemon shark; *Sphyrna tiburo*, bonnet shark; *Ginglymostoma cirratum*, nurse shark). Stimulation of central brain activity was marked in response to trimethylamine, betaine and trimethylamine oxide, whilst glutamic acid and tertiary amines were less excitatory.

Precisely which organs may be responsible for mediating the input to the brain that might bring about such responses are indicated by Bardach and Villars elsewhere in this volume. Nonetheless it is appropriate here to remark on the work of Katsuki *et al.* (1969) who showed that pit organs in *Ginglymostoma cirratum* do not appear to respond to organic substances, but rather act as salinity detectors, discharging in response to cations such as potassium, rubidium, sodium, ammonium, caesium and lithium. Divalent ions such as calcium and magnesium inhibit the discharge.

Attempts at more precise identification of stimulant substances have been made by Kleerekoper and Mogensen (1963) who found that *Petromyzon marinus* demonstrated sensitivity towards certain amines. The same kind of approach using electrophysiological methods was adopted by Case and Gwilliam (1963), Case (1964), Laverack (1963, 1964) and by Levandowsky and Hodgson (1965). The philosophy behind these experiments was a simple one, i.e. what substances may be found from time to time in the environment that may act as clues for feeding? Case concluded that glutamic acid was important; Laverack that trimethylamine oxide and betaine were significant for *Homarus* and *Carcinus*; Levandowsky and Hodgson showed that receptors responding to both types of compound were present in *Panulirus argus*, a finding also demonstrated by Laverack (1964) working on the same spiny lobster. These various substances are all found in fish and sundry other food of these decapods. It is probable in these cases that different receptor populations are responsible for different recorded responses.

Great specificity in the response of decapods to chemical stimuli has been shown in antennular preparations of the shrimp *Betaeus* (Ache and Case, 1969). *Betaeus hartfordi* and *B. macginitiae* are sensitive to at least two amino acids and other chemical substances released by other organisms, but notably to substances from the normal hosts of these commensal organisms. *B. hartfordi* responds to water from *Haliotis*; *Strongylocentrotus franciscanus* water stimulates *B. macginitiae*; that is to say water from the normal host stimulates the commensal.

B. Synergistic Action of Mixtures

As stated above electrophysiological investigations of isolated preparations from various parts of animals can at best give only a sketchy idea of what parameters are important to the animal in its daily life. The presentation of solutions of single amines, carbohydrates, lipids and other chemicals may demonstrate the sensitivity of certain cells towards these substances, but need not be very informative about their effectiveness as stimulants in the life of the animal.

Single chemical compounds stimulating the whole animal via an appropriate organ rarely seem to provoke the whole feeding response at least in marine forms, in flies things may be a little simpler (Dethier 1962). If some material that the animal normally feeds on is given certain usually stereotyped responses occur that are what we term feeding activities. If now single amines or other substances are presented in exactly the same way it is unusual if feeding responses occur. McLeese (1970) reporting on the American lobster *Homarus americanus* showed that some stimulation of the whole animal was obtained with organic chemicals (alanine, β-alanine, glutamic acid, proline, succinic acid, malic acid, and tyrosine), but only these 7 out of 29 tried were in any way effective. On the other hand 14 mixtures out of 15 tried were stimulatory. These solutions contained between 2 and 9 amino acids. Some evidence for interaction between these compounds was obtained, showing that responses to the sum of the individual components was greater or less than the activity expected from a consideration of the chemicals on their own. Whilst these experiments indicate that mixtures of this kind are closer to the normal situation at no time did such results mimic exactly the effects of normal stimulants such as fish, shrimp and lobster muscle.

Similar results using somewhat different compounds and mixtures have been shown for *Homarus gammarus* (Mackie and Shelton, personal communication) and *Carcinus maenas* (Shelton and Mackie, 1971). These synergistic effects may act at the periphery. In *Bombyx mori* strychnine stimulates the bitter receptor, but does so more markedly when salt is also present. Salt alone does not stimulate the bitter receptor (Ishikawa 1966). There are many possible permutations for this kind of interaction in chemosensory organs, but at present we cannot discriminate between the possibilities. Behavioural responses are at so many stages removed from the initial receptor site that only conjecture is possible.

VIII. Membrane Structure

It is probably time to abandon the apparent assumption that the membrane of the receptor cell is an unchanging structure which always presents the same face to the exterior. No matter whether one follows the classical interpretation of the Davson-Danielli bimolecular leaflet or some other structural appearance (such as that of Sjostrand) it must surely be considered as a dynamic structure. Ever since Schoenheimer (1942) and Rittenberg (1950) demonstrated that there is a dynamic turnover of intracellular substances the internal chemistry of the cell has been known to consist of eternally altering biochemical entities; the recent impetus in RNA synthesis in its various forms and for its various purposes also indicates a massive turnover of information and biochemical components. Why should not the membrane also be considered in the same way? A receptor site upon a membrane may well be synthesized within the cell, incorporated at the membrane as part of the lipid-protein layer, move through the thickness of the membrane and be exhibited at the surface there to remain (briefly?), or degrade and be cast off, or possibly reabsorbed into the cytoplasm of the cell for the components to be reutilized. From the point of view of synthesis of proteins this is the only way that substances can be incorporated into the membrane. With regard to turnover and substitution the work of Young and Bok (1969) and Young (1971) on the vertebrate rod has relevance. In this receptor the sensory membrane carrying the photo-sensitive pigments is supplied with raw material from the trophic cell body which passes proteinaceous substances via the ciliary region to the distal outer segment. These biochemical components then become part of the photo-receptive apparatus, gradually moving more and more distal to the cell body, and eventually cast off at the extreme tip of the receptor. The turnover time is 6–8 weeks. In this case perhaps the membrane is replaced as a whole by a repetitive cycle of biochemical synthesis on a basic template.

Such replacement may be the function of regrowth of taste bud cells mentioned previously. If the membrane cannot be resynthesized from within then shedding and restitution may provide an equivalent system. In cases where the cilia are dropped these may also be replaced by growth of new ones. But there are examples in which membrane

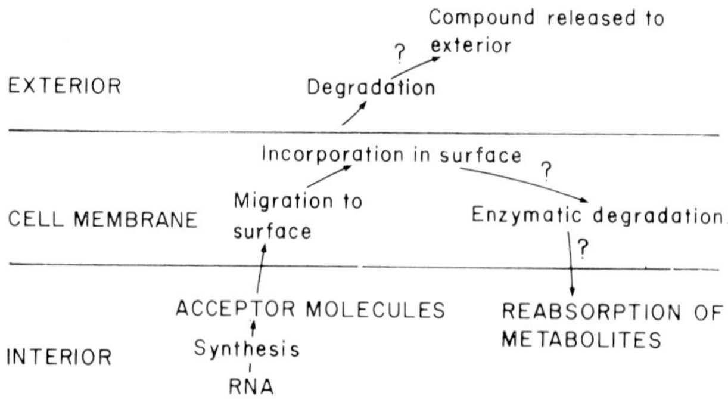

Fig. 5 A diagram to show possible relationships between the external environment, the cell membrane and the internal cellular region in terms of chemoreceptor/acceptor molecule organization. Molecules manufactured within the cell are incorporated in precise ways at particular places in the cell membrane, migrate to the external surface of that membrane and are incorporated into the structure at particular sites on the cell surface. They may then be degraded and the metabolites released into extracellular space, or undergo enzymatic alteration within the membrane and the products reabsorbed into the cell.

replacement must be a continuous phenomenon and may account for simple structural changes on a micro-scale (Fig. 5).

If this hypothesis is a reasonable one then it may mean that the cell surface is an ever changing one both in chemistry and molecular morphology since incorporation sites may vary from place to place at different times.

All sense organs adapt, some quickly others less so, in their electrical response to stimulation. The process of adaptation is a complex one, partly due to the physico/chemical properties of the sensory organ, partly to electrical changes of the membrane. In the case of the MRO mechanoreceptor system of crayfish Nakajima and Onodera (1969) have shown that about 70% of generator potential adaption is accounted for by the simple elastic-viscous property of the receptor muscle whilst adaptation of propagated potentials is a function of differences in the membrane. Especially important are the changes in permeability to the two ions Na^+ and K^+ and the electrogenic pump mechanism.

In the case of chemoreceptors adaptation may be an indication partly of blocking of acceptor sites by molecules already in place and not yet removed, and by membrane changes elsewhere. The rapid nature of removal of stimulants may account for the relatively slow adaptation of the receptors as electrophysiologically recorded, but a concerted attack by intracellular methods is necessary for a fuller understanding.

Isolation and characterization of specific proteins that might be concerned with chemoreception was reported by Dastoli, Lopieckes and Price (1968) and Dastoli, Lopieckes and Doig (1968). Mammalian taste bud proteins combined with substances of the bitter and sweet modalities, and hence were suggested as acceptor proteins. Koyama and Kurihara (1971) have refractionated taste bud proteins from cows and indicate that there is no difference between these and the surrounding epithelial cellular contents. They do not find evidence for specific acceptor proteins at the taste bud. On this basis these authors suggest either that acceptor proteins are distributed in all parts of the tongue; though possibly binding of stimulatory substances produces effective stimulation only at the taste bud, or alternatively the content of acceptor molecules in papillae is too low to be detected by the method used (polyacrylamide gel electrophoresis). If turnover rates are high this could conceivably be an explanation.

Enzyme kinetics and response curves of receptors indicates (Hansen, 1969) that enzyme substrate complexes occur, at least in insects (*Phormia*, sugar receptor). If a sugar glucosidase interaction takes place, one assumes that the enzyme responsible is a protein and so should be amenable to isolation and characterization by conventional means.

Despite the efforts mentioned above the biochemical nature of the chemoreceptor membrane is not known in detail at the present moment, but perhaps a sign has appeared on the horizon. It has long been argued that there are great similarities between postsynaptic and chemoreceptor membranes since both are affected by the action of chemicals, in a highly specific manner. It also seems that postsynaptic membranes may be arranged like a patchwork quilt with different acceptor sites built into the framework at different places, thus conferring sensitivity towards different transmitters at different areas on the cell (Dennis *et al.* 1971; Harris *et al.*, 1971; Kuffler *et al.*, 1971).

None of the prime biochemical agents in these responses have been identified until recently when Miledi *et al.* (1971) reported on the isolation of the cholinergic receptor protein of *Torpedo* electric tissue. By utilizing the irreversible specific blocker of acetylcholine, bungarotoxin (which combines with the acceptor protein specifically), the membrane fraction was separated from all other components, and some of its properties elucidated. The number of protein sites of this kind was estimated as $10{,}000/\mu^2$ of membrane. It may be some while before the

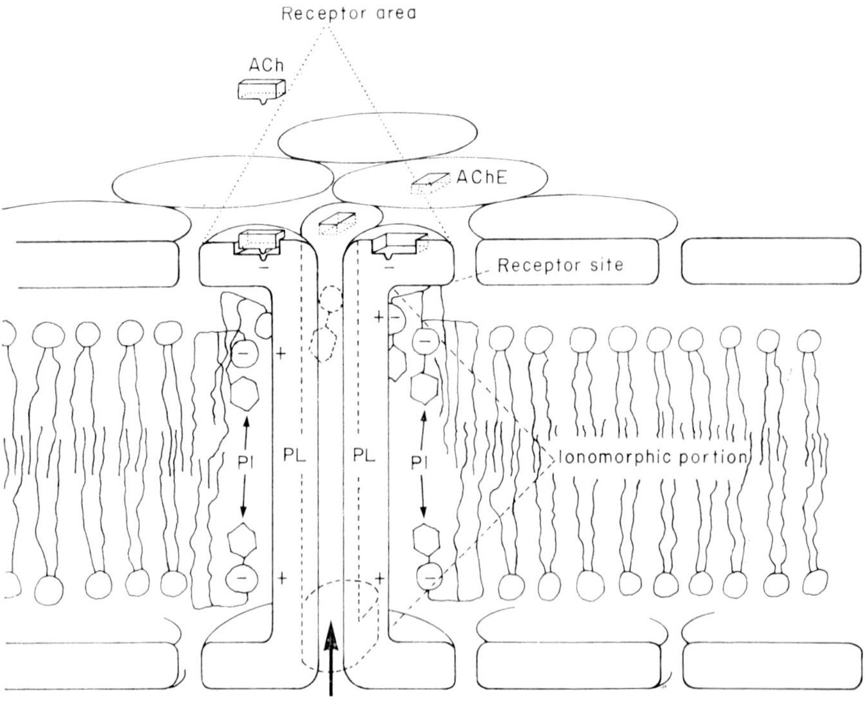

Fig. 6 The de Robertis reconstruction of a postulated postsynaptic membrane (analogous perhaps to a chemoreceptor membrane). Four proteolipid molecules (PL) in parallel traverse the membrane and with their ionophoric portions constitute the wall of a pore. An acceptor site (ACh) is shown at the outer end of the proteolipid. The acceptor area is constituted by the neighbouring acceptor sites and may form discrete patches on the cell surface. Acetyl-choline esterase as a separate macromolecule is shown also. This may not be appropriate for chemoreceptors, but other molecules may play a similar part. (From de Robertis, 1971, with permission of the Chicago University Press).

amino acid residues are analysed but the way is now open for an attack on similar problems.

A similar and parallel attack on the acceptor proteins of electroplax from *Torpedo* and *Electrophorus* has been reviewed by de Robertis (1971). His conclusions that a specific protein exists for the acceptance of acetylcholine at the postsynaptic membrane supports and extends the observations of Miledi *et al.* (1971). De Robertis gives a schematic view of what may happen at such a site (Fig. 6). It is conjectural, and by and large still only superficial since the mechanism by which a molecule at the surface affects membrane events in such a way as to allow ionic movements is still not clear. Changes in configuration seem to occur when the transmitter is present, and hence Duncan's (1967) hypothesis gains some further support, though it is not known at present that the isolated protein is in any way an ATPase. Nachmansohn (1970) also gives details of proteins isolated from excitable membranes. It so happens that electric organs are particularly valuable as sources of this particular protein and full advantage has been taken of this attribute. Nonetheless one looks forward to a similar attack on a chemoreceptor where acceptor proteins may be located. But first perhaps we need a specific blocker for a specific stimulant, and in this situation insect preparations may be important.

Alivastos *et al.* (1971) also threw light on methods of approach to the subject of interaction between stimulant and acceptor molecules. Using 5-hydroxytryptamine at a synaptic site they have shown that a chemical reaction may occur between the transmitter and the post-synaptic membrane. By study of Schiff bases it may be possible to isolate molecules from the cell membrane in order to identify their structure.

These authors also propose, in the way that I have attempted to follow here, that the term acceptor is more appropriate than receptor when applied to the processes of initial interaction of molecules. Receptors are those nerve cells that receive stimulation; acceptors are those parts of such cells that accept the stimulus (in the present case these are probably specific chemical substances). The term receptor has been widely used in this context, but is confusing, especially to receptor physiologists dealing with whole cells and not parts of cells. The value of the term acceptor is that it does not clash with other possible interpretations, and is specific in meaning and intention.

IX. Theories of Transduction

Information about the primary steps in stimulation of chemoreceptors is not easy to obtain since the first demonstrable physiological response may be at some remove from the first physico-chemical event at the cell surface. By virtue of considerable variety of response of cells to stimulation some aspects of the study become confusing. Nonetheless certain lines of approach seem more profitable than others and possibly a consensus of opinion now becomes available although there are differences that may yet prove to be crucial in the final analysis.

Access of the volatile or soluble substance to the cell surface seems to be largely a question of its solubility in water (especially in an aqueous protein medium like mucus), but perhaps also in some cases a question of solubility in lipid. Subsequent action upon the membrane may be related to the nature of that surface and the chemical reactivity. Thus the presence of enzymes or the morphology and configuration of the cell surface may be of overriding significance.

The various physico-chemical aspects of the problem have received considerable attention during the last decade or so, and certain preliminary conclusions may be suggested. That the morphology of the compound is important seems indicated by the work of Amoore and his colleagues (see for example Amoore *et al.*, 1969). Their research shows that the stereochemistry of chemicals is at least one important feature for stimulation of some types of chemoreceptor. Thus in the ant *Iridomyrmex pruinosus* 2-heptanone and other molecules with similar shape are stimulatory. In the case of the silk moth *Bombyx mori* of the four possible isomers only 10 *trans*-12-*cis*-hexadecadienol is of the right form for low threshold stimulation (Schneider, 1962). Schneider (1969) later reported that the sensitivity of the receptors to bombykol was such that one molecule was sufficient to stimulate one sense cell so the specificity and sensitivity are correspondingly great.

That the simple morphology and geometry of a molecule may not be enough on its own has been suggested by other workers notably Wright and Davies. Davies (1962) believes that the partition coefficient of substance between air–water and water–oil interface and the cross sectional area of the molecule are both important in "puncturing" the membrane and allowing ionic exchange. Wright (1964, 1966a, b) suggests that normal vibrational movements of molecules may play a part. Presumably in this case no molecule can exactly fit a particular hole in a membrane since the size will always be somewhat variable.

The possibility that specific chemicals are involved by incorporation into the cell membrane and play a part in transduction has led to a search for alterations in the biochemistry of membranes, that may reflect specific interactions. Davies (1962) thought that the number of specific chemicals necessary to explain all chemosensory phenomena was so vast as not to warrant consideration. Now with modern knowledge about the ability of cells to synthesize proteins from the normal metabolic raw materials the idea may not be so far fetched.

Ash (1968, 1969) has demonstrated certain absorbancy changes in preparations from rabbit olfactory mucosa. These alterations indicate a change in macromolecular conformation due to reactions between the stimulating substance and the cell components. It was not shown that this absorbancy change was restricted to membrane constituents, which is presumably necessary if this is a primary response to stimulation.

Later experiments showed that ascorbic acid is concerned in these reactions, being released from a chemically "bound" position when protein interacts with an external component. The response is not restricted to chemosensory cells, being found also in other tissue types. It thus seems to be a general response of cells to chemical stimulation (see also Edgar, 1970).

This work shows that cells may respond to chemical stimulation by biochemical events that are recordable but it may be a long way from the initial event that accepts a molecule at the cell surface and allows its translation into a change in membrane potential. In other words, the acceptor site or region may or may not be associated with ascorbic acid. It seems more likely that ascorbic acid is found in most types of cell and is mobilized when necessary.

Mozell (1970) has recently proposed a chromatographic model for the olfactory epithelium. In this hypothesis the nasal mucosa acts as a chromatogram or filter in which certain compounds may be absorbed from the tidal air differentially, thus removing some substances early in the air path, and others later. There is also a difference in the retention times of various substances. As a consequence of this it seems necessary to postulate that the cells at the nasal entrance may be of differing sensitivity to those elsewhere, and that there is a gradient of receptor cells of varying properties over the whole epithelial layer. There is a possibility that each individual receptor membrane acts as a chromatogram for the purpose of sorting out appropriate from inappropriate molecules, but this is not yet demonstrated.

Rosenberg *et al.* (1968) have again proposed a role for the carotenoid pigments which are noticeable in some olfactory systems. The occurrence of such pigments has previously been the subject of speculation and theories based on their presence have been criticized by Moulton (1962).

These criticisms may be listed as follows:

1. Not all olfactory epithelia possess carotenoid and other pigments (nor do taste buds and contact chemoreceptors).

2. That vitamin A deficiencies may affect responses and be reflected by anosmia, but only due to the common effect of vitamin A on the function of all membranes (cf. Vinnikov, 1965).

3. More satisfying theories take no account of, and are not dependent upon, the presence of carotenoids.

4. Pigment and olfactory areas do not always coincide.

There seems no reason in the light of recent work to modify these general criticisms.

Acceptor sites proper seem to be established in such systems as uterine mucosa (for hormone action) (Jungblut *et al.*, 1967) and on postsynaptic membrane (Miledi *et al.*, 1971) in a way that indicates specific protein molecules exquisitely adapted for their purpose. There seems to be some evidence therefore for the suggestion that by analogy chemoreceptor membranes are likewise organized.

Duncan (1967) proposed a theory of excitable cell properties that accounted for many isolated reports that have accumulated in the

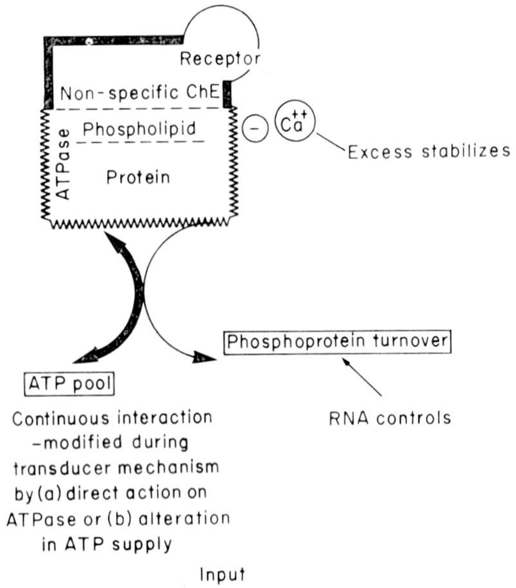

Fig. 7 The Duncan hypothesis is shown here as it applies to chemoreceptor cells. Actions of chemicals upon the cell membrane influence ATPase incorporated in the membrane. These enzymes are regenerated from within the cell, as also are the other components of the membrane. (Modified from Duncan, 1967.)

literature. The hypothesis briefly is that the membrane of the receptor carries a deformable ATPase mechanoenzyme that alters its molecular configuration when affected (by combination ?) by a stimulant molecule (see Fig. 7). The change in geometry allows ionic exchange to occur across the membrane and hence a receptor potential is set up. The reader is referred to Duncan's book for a full exposition.

I would only like to quote here one piece of work that seems to support Duncan's hypothesis. Kafka (1970) has carried out an exhaustive survey of the responses of receptors in sensilla coeloconica on the antennae of *Locusta migratoria*. A total of 370 separate compounds were tested and the most effective of these were unbranched keto-acids and unsaturated fatty acids with a chain length of 6 carbons. Unsaturated fatty acids and unsaturated alcohols were less stimulatory.

The site of action was localised as being a function of the dendritic membrane and molecules act as single individual entities without interaction with other molecules. The properties of such molecules are therefore important in determining their action. Analysis of all the responses shows that there is a reversible change of conformation in the acceptors of the membranes.

The general conclusions that might be reached at present therefore seem to be as follows:

1. Chemoreceptors have membranes that are adapted physiologically and structurally for the acceptance of external molecules.

2. The shape of the molecule is of significance in determining sensitivity.

3. The configuration of the membrane may be important as also may the charge upon it.

4. Acceptor molecules built into the membrane may be alterable in a way that mobilizes metabolic activity.

5. The transfer of ions across the membrane is a result of changes in the membrane structure, but the evidence available at the moment does not allow us to say how this happens.

It will be appreciated from this outline that most of the pertinent information on transduction has come from sources other than aquatic animals.

It should also be mentioned that many, if not all, sensory receptors have membranes electrically inexcitable at the transduction region. Thus although we may postulate ion flow through the membrane at the surface in response to stimulation, it is still necessary to account for transmission across the cell, by non-propagating graded means, and also for the amplification of the initial signal to the point where a propagated potential originates in the afferent axon (whether it be primary or secondary). The most likely candidate for internal transmitters would seem to be enzymatic mechanisms but little information is yet available.

Signal amplification is a topic dealt with as an intercellular, rather than (or perhaps as well as) intracellular, phenomenon, by Nicoll (1971) working on rabbits. Various forms of inhibition between receptor cells at the periphery (e.g. lateral inhibition in *Limulus* eye) have been postulated in the olfactory bulb (an intermediate relay station between receptors and the CNS). Dendrodendritic inhibition has been proposed (Rall and Shepherd, 1968), simply meaning that the cells of the olfactory bulb ramify and contact one another in such a way that mutual inhibition may occur around the focus of stimulated cell or cells. This would improve the signal to noise ratios of the system. Nicoll proposes that an amplification or increase in magnitude of the stimulus may also be effected by the process of monosynaptic interaction of an excitatory nature. A stimulated cell in its turn stimulates those around it via dendritic branches. Electrophysiological records of the activity of olfactory bulb cells reveals EPSPs driving spikes following electrical

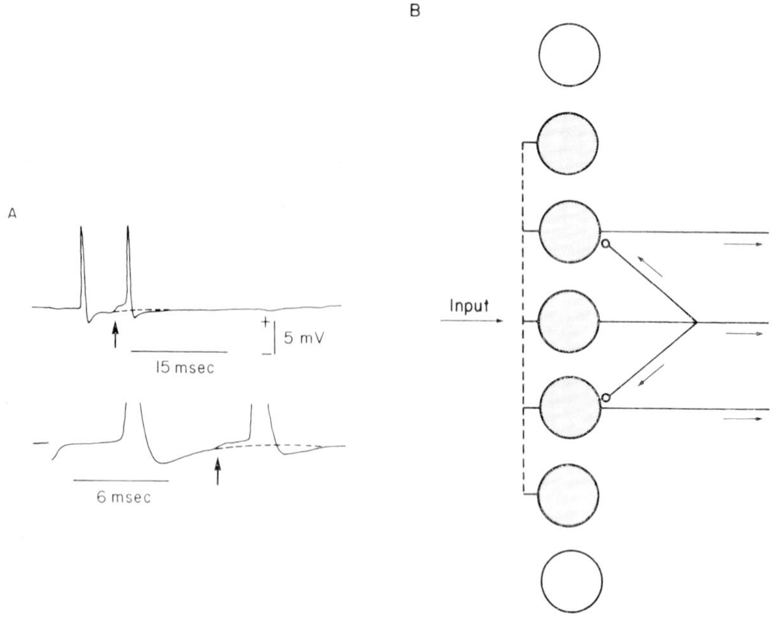

Fig. 8 (A.) Intracellular recording from a mitral cell of the olfactory bulb of a rabbit. The cell is responding to electrical stimulation along the lateral olfactory nerve (stimulus artefact precedes the first potential). Prior to the second spike a post-synaptic potential may be seen. The analysis of this type of record given by Nicoll (1971) is that there is direct synaptic connection between mitral cells within the bulb. (B.) This interaction is shown in diagrammatic form here. Ingoing signals affect mitral cells which in turn excite other neighbouring cells (via synaptic pathways). There is also an inhibitory surround which spreads rather further in the group (inhibited cells shown by shading). From Nicoll, 1971, by permission of the University of Chicago Press.

stimulation of the lateral olfactory tract (Fig. 8). There is therefore a direct link between units stimulated via this tract that give rise to EPSPs after a delay of about 7 msec. The immediate electrically stimulated response occurs within 1–2 msec and so a single synaptic delay is indicated.

Inhibition and excitation may take place in parallel as the result of natural stimulation and reinforcement of the afferent signal takes place at the bulb. This is of no consequence, of course, in other systems where the primary sensory neurones run direct without interaction to the CNS, as is the case in most invertebrates. Any amplification here must therefore be intracellular in nature.

X. Conclusions and Prospects

Lettvin and Gesteland (1965) seem to point to a possibly important difference between vertebrate and invertebrate olfactory systems. The majority of invertebrate chemosensory units, where characterized, fall into straightforward categories that can be labelled by their predominant response characteristics, though some units have a wider spectrum of response ("generalized" receptors; Schneider *et al.*, 1964).

Yamada (1971) working on the cockroach shows that odour coding at the olfactory lobe of the CNS may involve combinations of every olfactory neurone activity resulting in a unique cross-lobe pattern of discharges for each particular odorant. It would be possible to distinguish many odours by having an equivalent number of neurones possessing differential odour specificity. This is somewhat contrary to the viewpoint expressed by Gesteland *et al.* (1965) for vertebrates. In the frog olfactory system odours affect all sensory cells in some way. The very full analysis given by these authors will not be detailed here, but the reader is referred to the insight and the entertaining delivery of the general subject as given by Lettvin and Gesteland (1965). In this article they speculate about the nature of the response, and how several odours may have several effects on one cell, if one postulated the usually anticipated protein molecular sites on the membrane; and on other features summarized below (there are others not mentioned here).

1. Almost certainly there are several chemically sensitive molecules or molecular traps (probably many of them).

2. Either every receptor has many types of traps, or if only one trap per fibre, that one has a complex response to the odour environment. The response may be either excitatory or inhibitory, due either to ionic exchanges, or to transference of mechanical signals to the base of the cilium.

3. The receptor sees pure odours and combinations along a single dimension from most stimulatory to most depressant. If every fibre has a different point of view then each fibre represents a different dimension in a coordinate axis. Such a coding system preserved relations between odours in combination but the deciphering of the code remains inscrutable at the present point of time.

The implication of mechanical transmission along the length of the cilium suggested here is interesting in the light of Tucker's (1967) observation that the cilium can be removed without loss of the response, and to Matsusaka's (1967) report that ATPase activity exists in the ciliary rootlets of retinal rods (mammals) which he suggests may be involved in internal transmission.

Despite this résumé of the various phenomena that have been observed and described in the processes of chemical reception (Table 2)

Table 2. The properties of chemoreceptors. The features described in this review are tabulated here where it may be seen that two possibly distinct lines of evolution are found amongst chemoreceptor cells. Further information in the future will indicate whether this division is valid, or whether further additions must be made to this essentially simplistic picture

Type of receptor Dendritic ending	1 Ciliary	2 Microvillous	3 Rod shaped
Occurrence	Widespread olfactory Invertebrate systems	Taste buds some invertebrate examples	Fish
Physiological response	Active propagated potentials	Slow graded potentials	?
Numbers per organ	Often numerous	Usually in groups of small numbers	Few
Nature of nerve connection to CNS	Usually primary	Often secondary accessory fibre	
Efferent fibres	Not usually known	Some described	

it is probably true to state that we have as yet only the haziest idea of chemoreceptor activities. The many attempts and failures to classify odours and to categorize tastes even on psychophysical grounds, all tend to indicate how poor our understanding is at the moment. When one then moves on to the very specific systems of invertebrates then our knowledge is even less, but the *possibilities* may be greater.

We still as yet have no real understanding of:

1. The reaction between the chemical molecule and the membrane, though the work of Miledi *et al.* (1971) and de Robertis (1971) may be a guide in this,

2. Response of the membrane, which by analogy we propose to be similar to other excitable systems. What happens in the membrane to trigger the electrical events is not known.

3. The reason for the conductance changes observed, though these presumably reflect the membranous alterations.

4. How the resting situation is restored, i.e. how the odourous molecule is destroyed or released and the acceptor site returned to its original condition. Whether this is a simple reversible chemical reaction or whether the whole complex is autolysed and then rebuilt from precursors within the cells needs examination. If an acceptor molecule is incorporated into the membrane, how does it come to lie at the surface and not somewhere else within the membrane?

From this it can be realized that we have a long way to go before the phenomenon of chemoreception is properly understood.

Since this review was completed two important references have become available. Sato (1973) describes experiments relating to the responses of mammalian taste receptors, and the incorporation of specific receptor proteins at or near the cell membrane. Atema (1972) has proposed a new theory of transduction which places reliance on the presence of microtubules within a ciliary derived surface projection. He suggests that transduction occurs within the cell (not at the surface) and that incorporation into tubulin protein may be important. There is no indication as to what may happen in receptors lacking such microtubules.

References

Ache, B. and Case, J. (1969). An analysis of the antennular chemoreception in two commensal shrimps of the genus *Betaeus*. *Physiol. Zoöl.* **42**, 361–371.

Alivisatos, S. G. A., Ungar, F., Seth, P. K., Levitt, L. P., Geroulis, A. J. and Meyer, T. S. (1971). Receptors: localization and specificity of binding of serotonin in the central nervous system. *Science, N.Y.* **171**, 809–812.

Altner, H. and Boeckh, J. (1967). Uber das Reaktionsspektrum von Rezeptoren aus der Reichschleimhaut von Wasserfröschen (*Rana esculenta*). *Z. Physiol.* **55**, 299–306.

Altner, H. and Müller, W. (1968). Elektrophysiologische und elektronenmikroskopische Untersuchungen an der Reichschleimhaut des Jacobsonschen Organs von Eidechsen (*Lacerta*). *Z. vergl. Physiol.* **60**, 151–155.

Amoore, J. E., Palmieri, G., Wanke, E. and Blum, M. S. (1969). Ant alarm pheromone activity: correlation with molecular shape by scanning computer. *Science, N.Y.* **165**, 1266–1269.

Ash, K. O. (1968). Chemical sensing: an approach to biological molecular mechanisms using difference spectroscopy. *Science, N.Y.* **162**, 452–454.

Ash, K. O. (1969). Ascorbic acid: cofactor in rabbit olfactory preparations. *Science, N.Y.* **165**, 901–902.

Atema, J. (1972). Microtubule theory of sensory transduction. *J. Theor. Biol.* **38,** 181–190.

Bailey, D. F. and Laverack, M. S. (1966). Aspects of the neurophysiology of *Buccinum undatum* L. (Gastropoda). I Central responses to stimulation of the osphradium. *J. exp. Biol.* **44**, 131–148.

Bannister, L. H. (1965). The fine structure of the olfactory surface of teleostean fishes. *Q. J. microsc. Sci.* **106**, 333–342.

Bannister, L. H. (1968). Fine structure of the sensory endings in the vomeronasal organ of the slow-worm *Anguis fragilis*. *Nature, Lond.* **217**, 275–276.

Barber, V. C. and Wright, D. E. (1969). The fine structure of the sense organs of the cephalopod mollusc *Nautilus*. *Z. Zellforsch. mikrosk. Anat.* **102**, 293.

Barnes, D. J., Brauer, R. W. and Jordan, M. R. (1970). Triggering of the stomach eversion reflex of *Acanthaster planci* by coral extracts. *Nature, Lond.* **228**, 344–346.

Bayne, B. L. (1969). The gregarious behaviour of the larvae of *Ostrea edulis* L. at settlement. *J. mar. biol. Ass. U.K.* **49**, 327–356.

Beets, M. G. J. (1970). The molecular parameters of olfactory response. *Pharm. Rev., Lond.* **22**, 1–34.

Boeckh, J. (1967). Inhibition and excitation of single insect olfactory receptors, and their role as a primary sensory code. *In* "Second International Symposium on Olfaction and Taste" (T. Hayashi, ed.), pp. 721–735. Pergamon Press, London and New York.

Bullock, T. H. and Horridge, G. A. (1965). "Structure and Function in the Nervous Systems of Invertebrates". Freeman and Co., London.

Case, J. (1964). Properties of the dactyl chemoreceptors of *Cancer antennarius* Stimpson and *C. productus* Randall. *Biol. mar. biol. Lab. Woods Hole* **127**, 428–446.

Case, J. and Gwilliam, G. H. (1963). Amino acid detection by marine invertebrates. *Proc. XVI Internatn. Congr. Zool.* **3**, 75.

Castilla, J. C. and Crisp, D. J. (1970). Responses of *Asterias rubens* to olfactory stimuli. *J. mar. biol. Ass. U.K.* **50**, 829–847.

Chapman, H. D. and Wilson, R. A. (1970). The distribution and fine structure of the integumentary papillae of the cercaria of *Himasthla secunda* (Nicoll). *Parasitology* **61**, 219–227.

Cook, A., Bamford, O. S., Freeman, J. D. B. and Teideman, D. J. (1969). A study of the homing habit of the limpet. *Anim. Behav.* **17**, 330–339.

Cook, S. B. (1969). Experiments on homing in the limpet *Siphonaria normalis*. *Anim. Behav.* **17**, 679–682.

Crisp, D. J. and Meadows, P. S. (1962). The chemical basis of gregariousness in cirripedes. *Proc. R. Soc.* **B 156**, 500–520.

Crisp, M. (1971). Receptors and food location in a neogastropod. Ph.D. Thesis, University of Cambridge.

Dastoli, F. R., Lopiekes, D. V. and Doig, A. R. (1968). Bitter-sensitive protein from porcine taste buds. *Nature, Lond.* **218**, 884–885.

Dastoli, F. R., Lopiekes, D. V. and Price, S. (1968). A sweet-sensitive protein from bovine taste buds. Purification and partial characterization. *Biochemistry, N.Y.* **7**, 1160–1164.

Davies, J. T. (1962). The mechanism of olfaction. *Symp. Soc. exp. Biol.* **16**, 170–179.

Dennis, M. J., Harris, A. J. and Kuffler, S. W. (1971). Synaptic transmission and its duplication by focally applied acetylcholine in parasympathetic neurons in the heart of the frog. *Proc. R. Soc.* **B 177**, 509–539.

Dethier, V. G. (1962). Chemoreceptor mechanisms in insects. *Symp. Soc. exp. Biol.* **16**, 180–196.

Dethier, V. G. (1968). Chemosensory input and taste discrimination in the blowfly. *Science, N.Y.* **161**, 389–391.

Dethier, V. G. and Gelperin, A. (1967). Hyperphagia in the blowfly. *J. exp. Biol.* **47**, 191–200.

Dorsett, D. A. and Hyde, R. (1969). The fine structure of the compound sense organs of the cirri of *Nereis diversicolor*. *Z. Zellforsch. mikrosk. Anat.* **97**, 512–527.

Døving, K. B. (1966). Efferent influence upon the activity of single neurons in the olfactory bulb of the burbot. *J. Neurophysiol.* **29**, 675–683.

Døving, K. B. and Gemne, G. (1966). An electrophysiological study of the efferent olfactory system in the burbot. *J. Neurophysiol.* **29**, 665–674.

Duncan, C. J. (1967). "The Molecular Properties and Evolution of Excitable Cells". Pergamon Press, Oxford.

Easton, D. M. (1965). Impulses at the artefactual nerve end. *Cold Spring Harb. Symp. quant. Biol.* **30**, 15–28.

Edgar, J. A. (1970). Dehydroascorbic acid and cell division. *Nature, Lond.* **227**, 24–26.

Esakov, A. I. (1961). The efferent control of receptors (on the example of the chemoreceptors of the tongue). *Byull. eksp. Biol. Med.* **51**, 3–8.

Foster, B. A. and Nott, J. A. (1969). Sensory structures in the opercula of the barnacle *Elminius modestus*. *Mar. Biol.* **4**, 340–344.

Frazier, W. T., Kandel, E. R., Kupfermann, I., Waziri, R. and Coggeshall, R. E. (1967). Morphological and functional properties of identifiable neurons in the abdominal ganglion of *Aplysia californica*. *J. Neurophysiol.* **30**, 1288–1351.

Frings, H. and Frings, C. (1965). Chemosensory bases of food-finding and feeding in *Aplysia juliana* (Mollusca, Opisthobranchia). *Biol. Bull. mar. biol. Lab. Woods Hole.* **128**, 211–217.

Funke, W. (1968). Heimfindevermögen und Artstreue bei *Patella* L. (Gastropoda, Prosobranchia). *Oecologia* **2**, 19–142.

Galun, R. and Kindler, S. H. (1966). Chemical specifiicity of the feeding response in *Hirudo medicinalis* (L.). *Comp. Biochem. Physiol.* **17**, 69–73.

Gelperin, A. (1966). Investigations of a foregut receptor essential to taste threshold regulation. *J. Insect. Physiol.* **12**, 829–841.

Gemne, G. and Döving, K. B. (1969). Ultrastructural properties of primary olfactory neurons in fish (*Lota lota* L.). *Am. J. Anat.* **126**, 457–476.

Gesteland, R. C., Lettvin, J. Y. and Pitts, W. H. (1965). Chemical transmission in the nose of the frog. *J. Physiol., Lond.* **181**, 525–559.

Ghiradella, H., Case, J. and Cronshaw, J. (1968). Fine structure of the aesthetasc hairs of *Coenobita compressus* Edwards. *J. Morph.* **124**, 361–385.

Ghiradella, H., Cronshaw, J. and Case, J. (1968). Fine structure of the aesthetasc hairs of *Pagurus hirsutuisculus* Dana. *Protoplasma* **66**, 1–20.

Graziadei, P. P. C. (1964). Electron microscopy of some primary receptors in the sucker of *Octopus vulgaris*. *Z. Zellforsch. mikrosk. Anat.* **64**, 510–522.

Graziadei, P. P. C. and Bannister, L. H. (1967). Some observations on the fine structure of the olfactory epithelium in domestic duck. *Z. Zellforsch. mikrosk. Anat.* **80**, 220–228.

Graziadei, P. P. C. and Tucker, D. (1970). Vomeronasal receptor in turtles. *Z. Zellforsch. mikrosk. Anat.* **105**, 498–514.

Hansen, K. (1969). The mechanism of insect sugar reception, a biochemical investigation. *In* "Proceedings of the Third International Symposium on Olfaction and Taste" (C. Pfaffmann, ed.), pp. 382–391. Rockefeller University Press, New York.

Hara, T. J. (1970). An electrophysiological basis for olfactory discrimination in homing salmon: a review. *J. Fish. Res. Bd Can.* **27**, 565–586.

Harris, A. J., Kuffler, S. W. and Dennis, M. J. (1971). Differential chemosensitivity of synaptic and extrasynaptic areas on the neuronal surface membrane in parasympathetic neurons of the frog, tested by microapplication of acetylcholine. *Proc. R. Soc.* **B 177**, 541–553.

Hayes, W. F. (1971). Fine structure of the chemoreceptor sensillum in *Limulus*. *J. Morph.* **133**, 205–240.

Hendler, R. W. (1971). Biological membrane ultrastructure. *Physiol. Rev.* **51**, 66–97.

Hodgson, E. S., Mathewson, R. F. and Gilbert, P. W. (1967). Electroencephalographic studies of chemoreception in sharks. *In* "Sharks, Skates and Rays" (P. W. Gilbert, R. F. Mathewson and D. P. Rall, eds.), pp. 491–501. John Hopkins Press, Baltimore.

Horridge, G. A. (1963). Proprioceptors, bristle receptors, efferent sensory impulses, neurofibrils and number of axons in the parapodial nerve of the polychaete *Harmothoë*. *Proc. R. Soc.* **B 157**, 199–222.

Hughes, G. M. and Tauc, L. (1965). A unitary biphasic post-synaptic potential (BPSP) in *Aplysia*. *J. Physiol.* **179**, 27–28P.

Ishikawa, S. (1966). Electrical response and function of a bitter substance receptor associated with the maxillary sensilla of the larva of the silkworm *Bombyx mori* L. *J. cell. comp. Physiol.* **67**, 1–12.

Jahan-Parwar, B., Smith, M. and Baumgarten, R. von (1969). Activation of neurosecretory cells in *Aplysia* by osphradial stimulation. *Am. J. Physiol.* **216**, 1246–1257.

Jørgensen, C. B. (1955). Quantitative aspects of filter feeding in invertebrates. *Biol. Rev.* **30**, 391–454.

Jungblut, P. W., Hätzel, J., de Sombre, E. R. and Jensen, E. V. (1967). "Uber Hormonrezeptoren. Die Ostrogen bindenden Prinzipien der Erfolgsorgane". Springer, Heidelberg.

Kafka, W. A. (1970.) Molekulare Wechselwinkungen bei der Erregung einzelner Reichzellen. *Z. vergl. Physiol.* **70**, 105–143.

Katsuki, Y., Yanagisawa, K., Tester, A. L. and Kendall, J. I. (1969). Shark pit organs: response to chemicals. *Science, N.Y.* **163**, 405–407.

Kleerekoper, H. and Mogensen, J. (1963). Role of olfaction in the orientation of *Petromyzon marinus*. I. Response to a single amine in prey's body odor. *Physiol. Zoöl.* **36**, 347–360.

Koyama, N. and Kurihara, K. (1971). Do unique proteins exist in taste buds? *J. gen. Physiol.* **57**, 297–302.

Kuffler, S. W., Dennis, M. J. and Harris, A. J. (1971). The development of chemosensitivity in extrasynaptic areas of the neuronal surface after denervation of parasympathetic cells in the heart of the frog. *Proc. R. Soc.* **B 177**, 555–563.

Laverack, M. S. (1963). Aspects of chemoreception in crustacea. *Comp. Biochem. Physiol.* **8**, 141–151.

Laverack, M. S. (1964). The antennular sense organs of *Panulirus argus*. *Comp. Biochem. Physiol.* **13**, 301–321.

Laverack, M. S. (1968). On the receptors of marine invertebrates. *Oceanogr. mar. Biol. A. Rev.* **6**, 249–324.

Laverack, M. S. and Ardill, D. J. (1965). The innervation of the aesthetasc hairs of *Panulirus argus*. *Q. Jl. microsc. Sci.* **106**, 45–60.

Lawry, J. V. (1968). Efferent sensory impulses in annelids. *Comp. Biochem. Physiol.* **27**, 377–379.

Lawry, J. V. (1970). Mechanisms of locomotion in the polychaete *Harmothoë*. *Comp. Biochem. Physiol.* **37**, 167–179.

Lettvin, J. Y. and Gesteland, R. C. (1965). Speculations on smell. *Cold Spring Harb. Symp. quant. Biol.* **30**, 217–225.

• Levandowsky, M. and Hodgson, E. S. (1965). Amino acid and amine receptors of lobsters. *Comp. Biochem. Physiol.* **16**, 159–161.

Ling, E. A. (1969). The structure and function of the cephalic organ of a nemertine, *Lineus ruber*. *Tissue and Cell* **1**, 503–524.

Ling, E. A. (1970). Further investigations on the structure and function of cephalic organs of a nemertine *Lineus ruber*. *Tissue and Cell* **2**, 569–588.

Lo, S. E. and Acton, A. B. (1969). The ultrastructure of the rostral sensory organs of the water bug *Cenocorixa bifida* (Hungerford) (Hemiptera). *Can. J. Zool.* **47**, 717–722.

Lyons, K. M. (1969). Compound sensilla in monogenean skin parasites. *Parasitology* **59**, 625–636.

Matsusaka, T. (1967). ATPase activity in the ciliary rootlet of human retinal rods. *J. Cell. Biol.* **33**, 203–208.

McBride, J. R., Idler, D. R., Jonas, R. E. E. and Tomlinson, N. (1962). Olfactory perception in juvenile salmon. I. Observations on response of juvenile sockeye to extracts of food. *J. Fish. Res. Bd Can.* **19**, 327–334.

• McLeese, D. W. (1970). Detection of dissolved substances by the American lobster (*Homarus americanus*) and olfactory attraction between lobsters. *J. Fish. Res. Bd Can.* **27**, 1371–1378.

MacRae, K. (1967). The fine structure of sensory receptor processes in the auricular epithelium of the planarian, *Dugesia tigrina*. *Z. Zellforsch. mikrosk. Anat.* **82**, 479–494.

Miledi, R., Molinoff, P. and Potter, L. T. (1971). Isolation of the cholinergic receptor protein from *Torpedo* electric tissue. *Nature, Lond.* **229**, 554–557.

Moulton, D. G. (1962). Pigment and the olfactory mechanism. *Nature, Lond.* **195**, 1312–1313.

Mozell, M. M. (1970). Evidence for a chromatographic model of olfaction. *J. gen. Physiol.* **56**, 46–63.

Nachmansohn, D. (1970). Proteins in excitable membranes. *Science, N.Y.* **168**, 1059–1066.

Nakajima, S. and Onodera, K. (1969). Membrane properties of the stretch receptor neurones of crayfish with particular reference to mechanisms of sensory adaptation. *J. Physiol.* **200**, 161–185.

Nicoll, R. A. (1971). Recurrent excitation of secondary olfactory neurons: a possible mechanism for signal amplification. *Science, N.Y.* **171**, 824–826.

Nott, J. A. and Foster, B. A. (1969). On the structure of the antennular attachment organ of the cypris larva of *Balanus balanoides* L. *Phil. Trans. R. Soc.* **B 256**, 115–134.

Ong. J. E. (1969). The fine structure of the mandibular sensory receptors in the brackish water calanoid copepod (*Gladioferens pectinatus* Brauly). *Z. Zellforsch. mikrosk. Anat.* **97**, 178–195.

Rall, W. and Shepherd, G. M. (1968). Theoretical reconstruction of field potentials and dendrodendritic synaptic interactions in olfactory bulb. *J. Neurophysiol.* **31**, 884–915.

Reese, T. S. (1965). Olfactory cilia in the frog. *J. Cell Biol.* **25**,(2), 209–230.

Rittenberg, D. (1950). Dynamic aspects of the metabolism of amino acids. *Harvey Lectures* **44**, 200–219.

deRobertis, E. (1971). Molecular biology of synaptic receptors. *Science, N.Y.* **171**, 963–971.

Rosenberg, B., Misra, T. N. and Switzer, R. (1968). Mechanism of olfactory transduction. *Nature, Lond.* **217**, 423–427.

Ross, M. M. R. (1967). Modified cilia in sensory organs of juvenile stages of a parasitic nematode. *Science, N.Y.* **156**, 1494–1495.

Sandeman, D. C. (1965). Electrical activity in the radial nerve cord and ampillae of sea urchins. *J. exp. Biol.* **43**, 247–256.

Sato, M. (1973). Gustatory receptor mechanism in mammals. *Adv. in Biophysics* **4**, 103–152.

Sato, T. (1969). The response of frog taste cells (*Rana nigromaculata* and *Rana catesbeana*). *Experientia* **25**, 709–710.

Schneider, D. (1962). Electrophysiological investigations on the olfactory specificity of sexual attracting substances in different species of moths. *J. Insect Physiol.* **8**, 15–30.

Schneider, D. (1969). Insect olfaction: deciphering system for chemical messages. *Science N.Y.* **163**, 1031–1037.

Schneider, D., Lacher, V. and Kaissling, K. E. (1964). Der Reaktionweise und das Reaktionsspektrum von Reichzellen bei *Antharae pernyii* (Lepidoptera, Saturiidae). *Z. vergl. Physiol.* **48**, 632–662.

Schoenheimer, R. (1942). "The Dynamic State of Body Constituents". Harvard University Press, Cambridge, Massachusetts.

Shelton, R. G. J. and Laverack, M. S. (1970). Receptor hair structure and function in the lobster, *Homarus gammarus* (L.). *J. exp. mar. Biol. Ecol.* **4**, 201–210.

Shelton, R. G. J. and Mackie, A. M. (1971). Studies on the chemical preferences of the shore crab, *Carcinus maenas* (L.). *J. exp. mar. Biol. Ecol.* **7**, 41–49.

Sjostrand, F. S. and Barajas, L. (1968). Effect of modifications in conformation of protein molecules on structure of mitochondrial membranes. *J. Ultrastruct. Res.* **25**, 121–155.

Snyder, N. and Snyder, H. (1970). Alarm response of *Diadema antillarum*. *Science, N.Y.* **168**, 276–278.

Steinbrecht, R. A. (1969a). On the question of nervous syncytia: lack of axon fusion in two insect sensory nerves. *J. Cell Sci.* **4**, 39–53.

Steinbrecht, R. A. (1969b). Comparative morphology of olfactory receptors. *In* "Third International Symposium on Olfaction and Taste" (C. Pfaffmann, ed.), pp. 3–21. Rockefeller University Press, New York.

Stinnakre, J. and Tauc, L. (1969). Central neuronal response to the activation of osmoreceptors in the osphradium of *Aplysia*. *J. exp. Biol.* **51**, 347–361.

Stockenius, W. and Engelman, D. M. (1969). Current models for the structure of biological membranes. *J. Cell Biol.* **42**, 613–646.

Storch, V. and Welsch, U. (1969a). Uber Bau und Funktion der Nudibranchie Rhinophoren. *Z. Zellforsch. mikrosk Anat.* **97**, 528–536.

Storch, V. and Welsch, U. (1969b). Zur Feinstruktur des Nuchalorgans von *Eurythoë complanata* (Pallas) (Amphinomidae, Polychaetae). *Z. Zellforsch. mikrosk. Anat.* **100**, 411–420.

Szal, R. (1971) "New" sense organ of primitive gastropods. *Nature, Lond.* **229**, 490–492.

Thomas, W. J. (1970). The setae of *Austropotamobius pallipes* (Crustacea: Astacidae). *J. Zool., Lond.* **160**, 91–142.

Thornhill, R. A. (1967). The ultrastructure of the olfactory epithelium of the lamprey *Lampetra fluviatilis*. *J. Cell Sci.* **2**, 591–602.

Thornhill, R. A. (1970). Cell division in the olfactory epithelium of the lamprey *Lampetra fluviatilis*. *Z. Zellforsch. mikrosk. Anat.* **109**, 147–157.

Thurm, U. (1968). Steps in the transducer process of mechanoreceptors. *Symp. zool. Soc. Lond.* **23**, 199–216.

Tucker, D. (1967). Olfactory cilia are not required for receptor function. *Fedn. Proc. Fedn. Am. Socs. exp. Biol.* **26**, 544, Abs. no. 1609.

Vinnikov, J. A. (1965). Principles of structural, chemical and functional organization of sensory receptors. *Cold Spring Harb. Symp. quant. Biol.* **30**, 293–299.

Welsch, U. and Storch, V. (1969a). Uber das Osphradium der prosobranchen Schnecken *Buccinum undatum* L. and *Neptunea antiqua* (L.). *Z. Zellforsch. mikrosk. Anat.* **95**, 317–330.

Welsch, U. and Storch, V. (1969b). Die Feinstruktur der Geschmacksnospen von Welsen *Clarias batrachus* L. und *Kryptopterus bicirrhis* (Cuvier et Valenciennes). *Z. Zellforsch. mikrosk. Anat.* **100**, 552–559.

Whitear, M. (1965). Presumed sensory cells in fish epidermis. *Nature, Lond.* **208**, 703–704.

Whitear, M. (1971a). The free nerve endings in fish epidermis, *J. Zool., Lond.* **163**, 231–236.

Whitear, M. (1971b). Cell specialization and sensory function in fish epidermis. *J. Zool., Lond.* **163**, 237–264.

Willows, A. O. D. (1971). Giant brain cells in molluscs. *Scient. Am.* **224** (**2**), 68–75.

Wilson, D. P. (1968). The settlement behaviour of the larvae of *Sabellaria alveolata*. *J. mar. biol. Ass. U.K.* **48**, 387–435.

Wilson, J. A. F. and Westerman, R. A. (1967). The fine structure of the olfactory mucosa and nerve in the teleost *Carassius carassius* L. *Z. Zellforsch. mikrosk. Anat.* **83**, 196–206.

Wolbarsht, M. L. and Hanson, F. E. (1967). Electrical and behavioral responses to amino acid stimulation in the blowfly. *In* "Proceedings of the Second International Symposium on Olfaction and Taste" (T. Hayashi, ed.), pp. 749–760. Pergamon Press, London and New York.

Wood, L. (1968). Physiological and ecological aspects of prey selection by the marine gastropod *Urosalpinx cinarea* (Prosobranchia: Muricidae). *Malacologia* **6**, 267–320.

Wright, R. H. (1964). "The Science of Smell", 164 pp. Allen and Unwin, London.

Wright, R. H. (1966a). Why is an odour? *Nature, Lond.* **209**, 551–554.

Wright, R. H. (1966b). Odour and molecular vibration. *Nature, Lond.* **209**, 571–573.

Yamada, M. (1971). A search for odour encoding in the olfactory lobe. *J. Physiol.* **214**, 127–143.

Yates, R. D., Chen, I.-L. and Duncan, D. (1970). Effects of sinus nerve stimulation on carotid body glomus cells. *J. Cell Biol.* **46**, 544–552.

Young, R. W. (1971). Shedding of discs from rod outer segments in the Rhesus monkey. *J. Ultrastruct. Res.* **34**, 190–203.

Young, R. W. and Bok, D. (1969). Participation of the retinal pigment epithelium in the rod outer segment renewal process. *J. Cell Biol.* **42**, 392–403.

2. The Chemical Senses of Fishes*

J. E. BARDACH

Hawaii Institute of Marine Biology, University of Hawaii, Honolulu

and

TRUDY VILLARS

Department of Psychology, University of Michigan, Ann Arbor, Michigan, U.S.A.

I. Introduction

In this chapter the anatomy of fish chemoreceptive systems and their capabilities will be briefly described. The role of the chemical senses in the behavior of fishes will be discussed as well as deficiencies in the information presently available and the need for future research.

The basic mechanism of the receptor function of smell and taste in all vertebrates is less understood than sight, hearing, touch and the lateral line. However, the importance of the chemical senses in the lives

*A substantial portion of the work described here was supported by U.S. Public Health Service Grants NS-04867; contribution 428 of the Hawaii Institute of Marine Biology.

of aquatic animals becomes more apparent each year. Among fishes, taste is restricted primarily to aspects of food finding and feeding, but smell, while it may serve for food detection and location, is also important in reproductive and social behavior.

Fish can smell volatile substances in great dilution, such as oil of cloves or phenols and their derivatives. They smell and taste various amino acids (Sutterlin and Sutterlin, 1971; Bardach, *et al.*, 1967a) thus demonstrating some functional overlap between their olfactory and gustatory domains. Various reactions to specific smells such as to home streams, home sites, species mates, enemies, reproductive partners, and young have been demonstrated but the nature of the compound or compounds involved in these reactions remains unknown.

Differences between smell and taste are both physiological and anatomical. Smell responses at their most acute levels can be triggered by a few molecules in a fish's nasal chamber (Teichmann, 1962); while taste thresholds of fishes, far lower in certain species than those of man (Glaser, 1966), are in the micromolar range and to be effective require millions of molecules per cm^3 of a stimulant to bathe the animal's taste sensors on its body or in its mouth.

Smell and taste are also clearly distinguished by the anatomy of their sensors. Smell sensors are the ciliated receptor cells in the olfactory epithelium on the lamellae in the nasal pit (Holl, 1965), whereas the taste cells have their microvilli gathered in taste buds and spread over the body, concentrated on filamentous fins or barbels and/or distributed selectively in oral and gill cavities. The nerves that connect these sensory transducers to the brain are in the first cranial or olfactory nerve and the taste fibers of cranial nerves VII, IX, and X, respectively.

Psychophysical tests in man have established the four basic taste qualities and furnished the description of smell categories such as fruity, minty, etc. Since such descriptions cannot be obtained from animals, the capabilities of their chemoreceptors are best described by listing the kinds of compounds they detect and by testing for specificity of receptors and/or nerves. For instance, Konishi and Zotterman (1963) found seven taste fiber types in the carp which react either to single taste substances or to a combination of them. Konishi, *et al.*, (1966), observed some taste fibers in the sea catfish (*Plotosus*) reacting to a specific stimulus emitted by a prey; he called them *Nereis* fibers. Also tasted are some larger compounds, customarily not associated with taste, such as glycerophospholipids, (Konishi and Zotterman, 1963; Bardach, *et al.*, 1967a) and nucleotides (Kiyohara *et al.*, in press).

A common chemical sense has been postulated, especially for aquatic animals, by some workers (Parker, 1922); its existence has been denied by others (Buddenbrock, 1952). It may rely for its functioning on numerous free nerve endings in the epidermis which are suggested to subserve pain sensation by Whitear (1971a), a function not inconsistent with the perception of some chemical stimuli. The recent investigations also by Whitear (1965, 1971b) indicate that a "common chemical sense" in fishes, at least, could rely altogether or in part on small single chemosensory cells in the epidermis which are spinally innervated. It could also rely on the chemical sensitivity of lateral line receptors. Katsuki, *et al.*, (1971a and b) have shown these to be responsive to monovalent cations, especially potassium, and divalent cations aside from responding to mechanical stimuli. Their experimental results suggest the free standing and pit organs of the lateral line system to have the dual functions of mechano and chemoreception. The electrophysiological evidence they offer is circumstantial and while persuasive, it remains to be demonstrated behaviorally whether or not fishes, amphibian larvae and aquatic amphibians actually use the above mentioned lateral line elements as chemosensory organs.

It is possible that fishes possess as many as five different chemosensory systems: (1) olfaction; (2) taste buds consisting of several taste cells which are cranially innervated sensors, in the mouth and in many species also on their bodies; (3) single taste cells on their bodies as well as perhaps in their oral cavities which are spinally innervated and may have a lesser gamut of sensitivities than the taste buds; (4) free nerve endings; and (5) the lateral line system, which has a limited range of capabilities to sense chemicals. As less is known about free nerve ending and lateral line chemosensory systems than about olfactory and taste bud receptors, they have been mentioned in this introduction for completeness' sake but they will not be discussed any further.

II. Anatomy of Chemosensory Structures

A. Smell

1. *The Nose*

In order to bring dissolved or dispersed stimulants to the olfactory receptors in the nasal pit, which has no connection with the mouth (except in lungfishes), certain mechanisms have evolved among fishes that serve to transport water through the nose.

A stream of water may be led to a fish's nose through a groove enhanced by the configuration of the head, as in the hammerhead shark (*Sphyrna*), and in many rays, as in the guitarfish (*Rhinobatos*) (Fig. 1). A flap of skin and/or connective tissue is further instrumental in funneling the water through the relatively open nasal depression with its

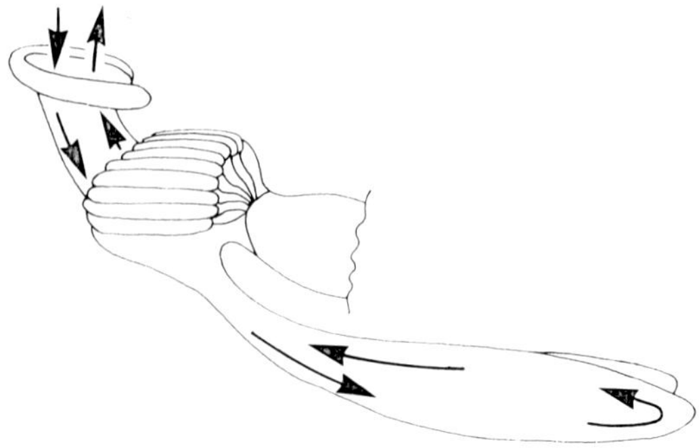

(a)

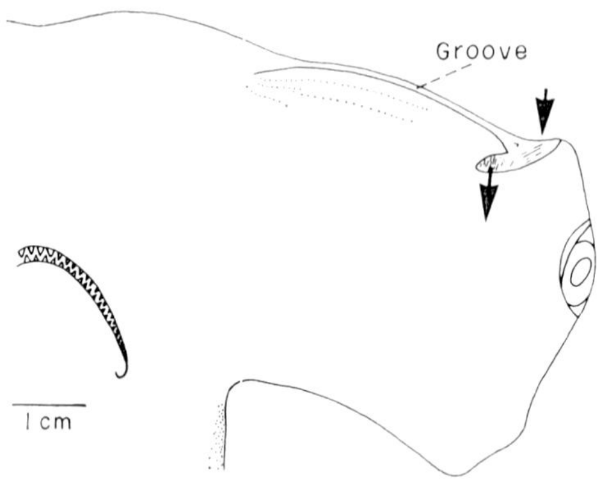

(b)

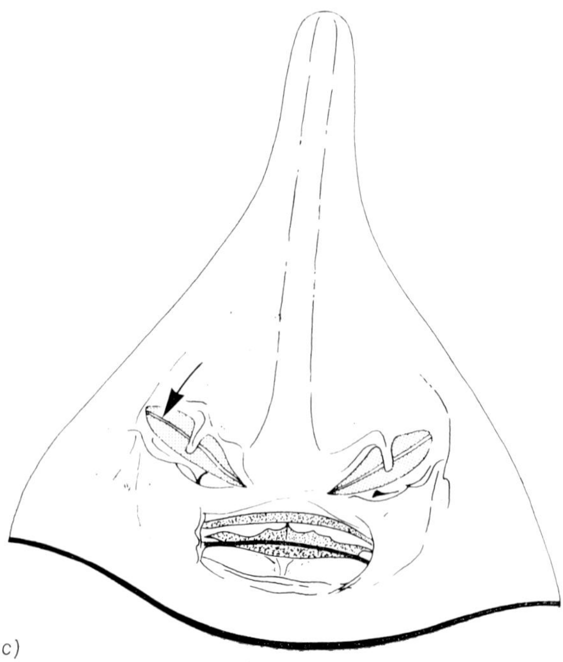

(c)

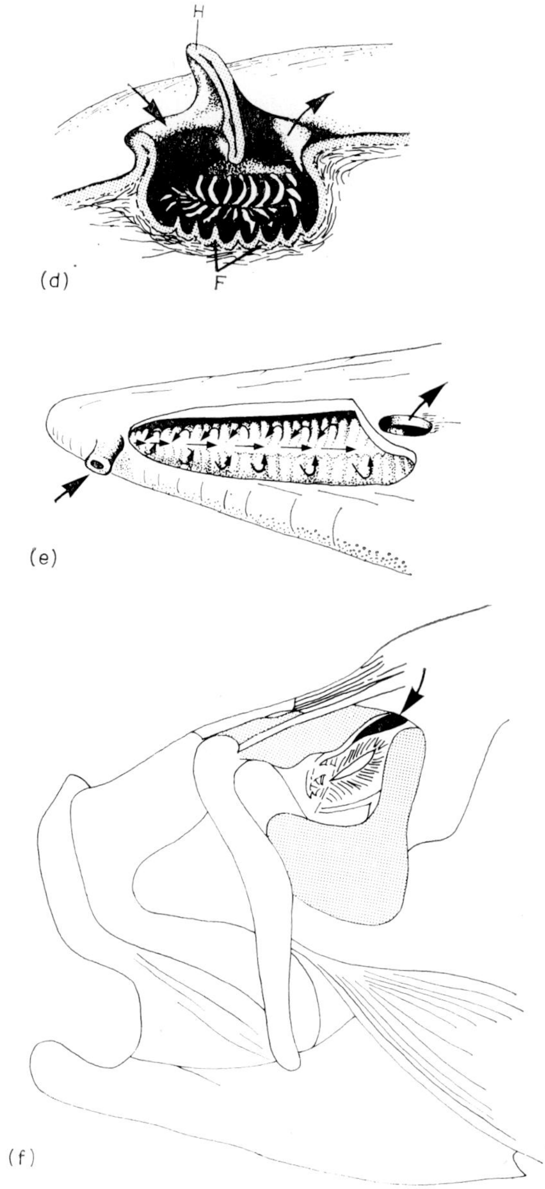

Fig. 1 Gross anatomy of the fish nose indicating mechanism for the perfusion of nasal chamber. (a) Sea Lamprey (*Petromyzon marinus*) note single opening, folds of mucosa and nasopharyngeal pouch; after Kleerekoper and van Erkel (1960). (b) Hammerhead Shark (*Sphyrna*) after Tester (1963). (c) Guitarfish (*Rhinobatos*) after Kleerekoper (1969). (d) Minnow (*Phoxinus*) after Teichmann (1962). (e) Eel (*Anguilla*) after Teichmann (1962). (f) Mullet (*Mugil*) after Burne cited by Kleerekoper (1969).

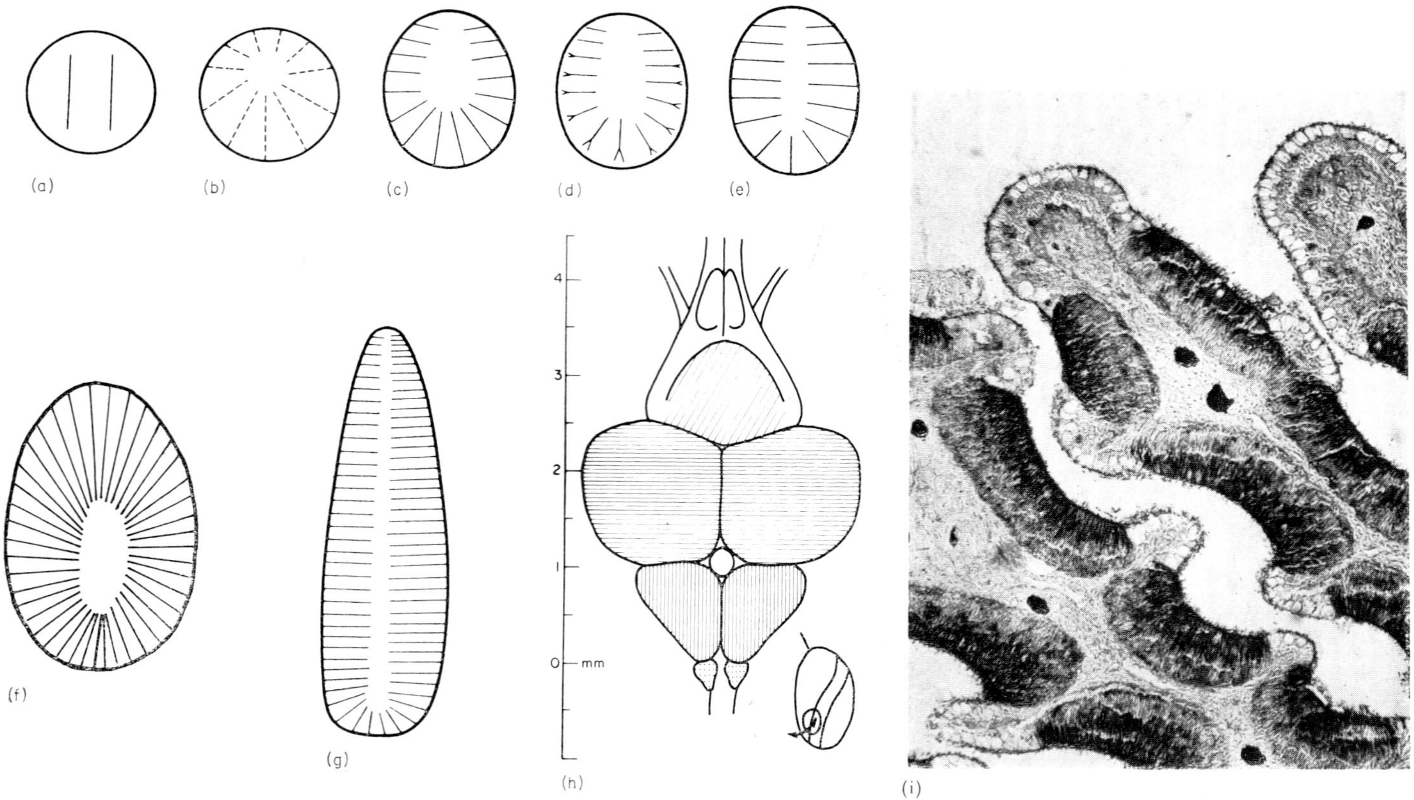
(a)
(b)
(c)
(d)
(e)
(f)
(g)
4
3
2
1
0
mm
(h)
(i)

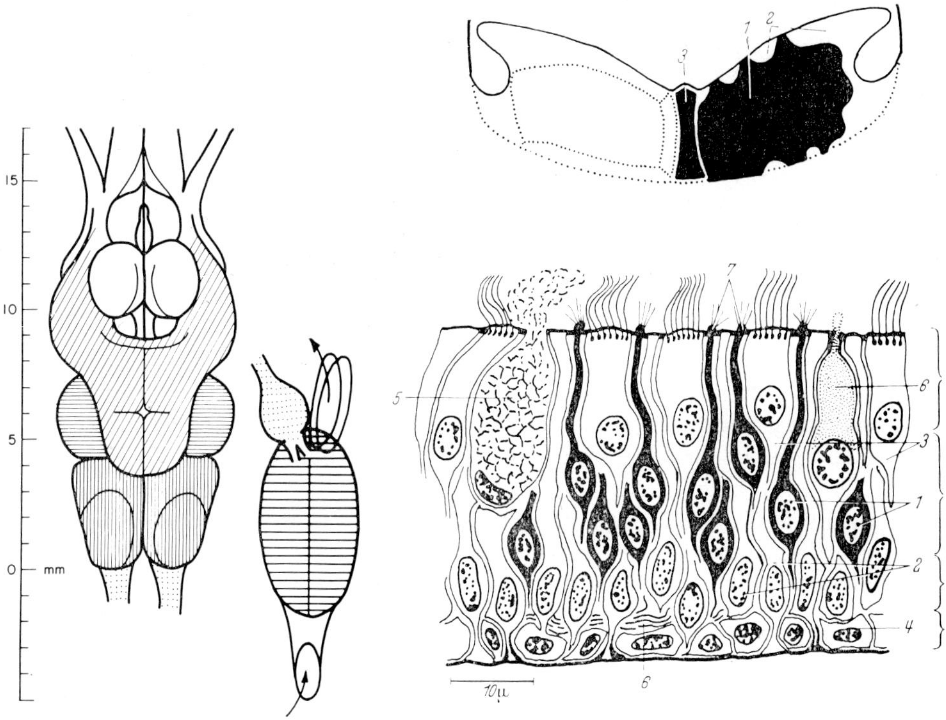

(j) (k)

Fig. 2 Aspects of fish olfactory organs.
(a–g) Schematic representations of numbers and arrangements of lamellae in olfactory rosette after Teichmann (1954). (a) Stickleback (*Gasterosteus*) (b) Pike (*Esox*) (c) Rainbow Trout (*Salmo*) (d) Perch (*Perca*) (e) Minnow (*Phoxinus*) (f) Bullhead (*Ictalurus*) (g) Eel (*Anguilla*).
(h + j) Brains of Stickleback and Bullhead respectively showing sizes of olfactory (bottom) as opposed to optic portions (center). Note relatively large optic lobe in h and a large olfactory one in j and olfactory pits with only two lamellae in h and many in j.
(i) Planar section through olfactory folds of a salmon; showing fields of olfactory cells (Bodian silverstain). (k) Distribution of cell types on an olfactory lamella of the eel (from Holl, 1965). (Reproduced with permission from Springer–Verlag.)
Top (1) olfactory epithelium (2) indifferent epithelium (3) median raphe. Bottom. Detail of area, above (1) receptors (2) supporting cells (3) ciliated cells (4) basal cells (5) goblet cell (6) club shaped secretory cells (7) olfactory vesicle or cell with sensory hairs.

olfactory epithelium arranged in folds or lamellae. This is again found in elasmobranchs as well as among a great many teleosts such as the Elritze (*Phoxinus*) (Fig. 1) and other minnows and the pike (*Esox*).

Other fishes employ hydraulic devices: A sac (as in the lamprey, *Petromyzon*, Fig. 1) or sacs are filled and emptied by the respiratory movements of head and branchial skeleton so that water is driven over the olfactory folds. The anatomy of such hydraulic exchange mechanisms was described by Burne (1909) but their flow dynamics are not well understood. One may suspect an incomplete exchange of water to occur in many cases and with it a stimulus may linger or be repeated beyond its significant time span. Such a contingency would have to be dealt with by central mechanisms and feedback organization. These are likely to come into play because elaborate integration provisions exist in the olfactory bulb and antidromic impulses have been observed in the olfactory tract.

Fishes with the best developed sense of smell have the most numerous olfactory folds; the eel (*Anguilla*) and the bullhead (*Ictalurus*) (Fig. 2, f and g) are pertinent examples. The elongated, closed nasal pits may have moveable intake tubes (eel). Water is drawn through the nose by an intricately patterned movement of cilia (Fig. 1). The flow can be speeded up or slowed down either in both nasal chambers or in one alone, as in the bullhead. Intake tubes that can be moved voluntarily and independent ciliary control of one or the other nasal chamber may serve to enhance detection of stimulus concentration differences on the right as opposed to the left side of the fish. It should be noted that there are cilia on the supporting cells of every fish nasal epithelium; these help to propel the stimulus-carrying boundary layer of water over the mucus-covered sense cells. The role of the mucus in the transduction of the stimulus is unknown.

The skipjack tuna (*Katsuwonus pelamis*), a fast swimmer that strongly reacts to food smells (Bardach and Todd, 1970), employs all three mechanisms (as do many fishes) for propelling the water through the nose: there is a groove on each side of the forehead, leading to the incurrent naris; cilia create a vortex pattern over the lamellae and hydraulic sacs assist in the ventilation of the nose (Gooding, 1963).

The elaboration of various nasal mechanisms such as moveable nasal intake tubes and a large number of olfactory folds are certain measures of olfactory acuity. Compare, for instance, the nose of the nearly anosmic sticklebacks (*Gasterosteidae*) (Fig. 2) with that of the bullhead or the eel (Fig. 2). The relative importance of the various senses in a fish's life is further illustrated by a look at the relative sizes of the various centers in the brain of the stickleback and bullhead, respectively (Fig. 2, bottom). Atema, *et al.*, (1969), have suggested, on the basis of behavior observations of these two animals, that fishes with a well developed smell sense have more variable and "plastic" behavior than those that depend primarily on sight.

Aside from the number of folds in the nose, the arrangement of sensory in relation to supporting epithelium also differs among different fishes. As one would expect, the more acute the animal's sense of smell, the larger the areas covered with sensory as opposed to supporting epithelium (Teichmann, 1954). In the pike (*Esox*) for instance, a fish that does not rely greatly on smell, the folds are low and few and there are sparse, small circular patches of sensory tissue on and between them. In the salmon (Fig. 2), a silver stained planar section of the nose shows numerous dark islets of olfactory cells among hillocks of mucus secreting supporting ells. Figure 2, depicting the olfactory fold of an eel, shows that mucus as well as supporting cells occur in the patches of olfactory epithelium. The ultrastructure of fish olfactory cells is reviewed in Kleerekoper (1969); its treatment is outside the realm of this article.

2. *The Olfactory Bulb*

The vertebrate sense of smell, like sight, is characterized by peripheral integration and convergence, that is by a reduction in numbers of higher order conducting neurons in the olfactory tract as compared to the number of axons of the primary or sensory neurons in the olfactory nerve proper that lead to the olfactory bulb. For the burbot, for instance, Kleerekoper (1969) quotes Døving's estimate of 10^7 olfactory cell axons leading to the bulb where they become reduced a thousand fold to provide the tract with about 10^4 fibers. Great variety exists in the topography of fish brains, especially in the spatial relation of the bulb to the brain and the nose, respectively; some fishes have long, others short olfactory nerves, with corresponding differences also in the lengths of their olfactory tracts. Figure 3, prepared by Professor Uchihashi, illustrates these variations for Japanese fishes. Such information is of value to neurophysiologists who may seek easy access to primary or higher order smell neurons. The figure also supports the suggestion by Aronson and Kaplan (1968) that a long olfactory tract as opposed to a long olfactory nerve permits faster conduction of olfactory information. Such an arrangement would benefit fishes with a well developed taste system on their bodies because it would permit better integration and coordination of impulses received in the two systems; the long tracts among the cyprinid and gadid fishes—both with many external taste buds—are cases in point (Fig. 3).

The neuro-anatomy of the olfactory bulb of fishes resembles that of other vertebrates. From outward to inward there is a layer of olfactory fibers, and then a layer where these interlace with dendritic structures

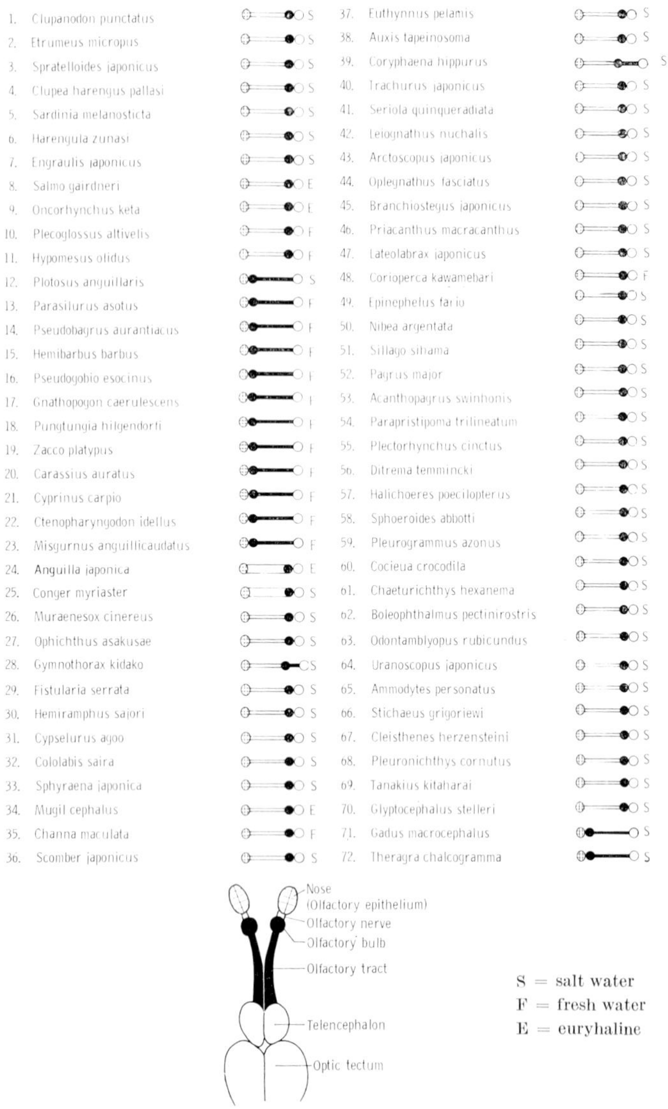

Fig. 3 The relative lengths of olfactory nerves and tracts in certain Japanese fishes, after Uchihashi (1953). List of Families, by numbers; 1–6 Clupeidae; 7 Engraulidae; 8, 9 Salmonidae; 10 Plecoglossidae; 11 Osmeridae; 12 Plotosidae; 13 Siluridae; 14 Bagridae; 15–22 Cyprinidae; 23 Cobitidae; 24 Anguillidae; 25 Congridae; 26 Muraenesocidae; 27 Ophichthidae; 28 Muraenidae; Fistularidae; 30 Hemiramphidae; 31 Exocoetidae; 32 Scombresocidae; 33 Sphyraenidae; 34 Mugilidae; 35 Channidae; 36–38 Scombridae; 39 Coryphaenidae; 40, 41 Carangidae; 42 Leiognathidae; 43 Trichodontidae; 44 Oplegnathidae; 45 Branchiostegidae; 46 Priacanthidae; 47, 48 Percichthyidae; 49 Serranidae; 50 Sciaenidae; 51 Sillaginidae; 52, 53 Sparidae; 54–55 Pomadasidae; 56 Embiotocidae; 57 Labridae; 58 Tetrodontidae; 59 Hexagrammidae; 60 Platycephalidae; 61, 62 Gobiidae; 63 Gobioididae; 64 Uranoscopidae; 65 Ammodytidae; 66 Stichaeidae; 67–70 Pleuronectidae; 71, 72 Gadidae.

of mitral cells in the distinct aggregations of the glomeruli, and finally layers of external and internal cells from which arise the olfactory tract fibers. Distinction is possible between afferent and efferent fibers in the bulb. In the center of the bulb there is an olfactory ventricle.

Detailed functional analysis of the fish olfactory bulb is lacking; it is assumed to "sort out" stimuli and to begin to categorize them for further elaboration in the brain. In fact, Bruckmoser (1973) suggests that the bulb processes the bulk of the olfactory signals. It is also likely to be an important way station in the regulation of primary neuron function through antidromic impulses. The latter has been discussed by Døving (1966), and Hara and Gorbman (1967).

3. *Olfactory Brain Centers*

The olfactory bulb connects with the forebrain through the olfactory tract, which consists of lateral and medial bundles. The separation

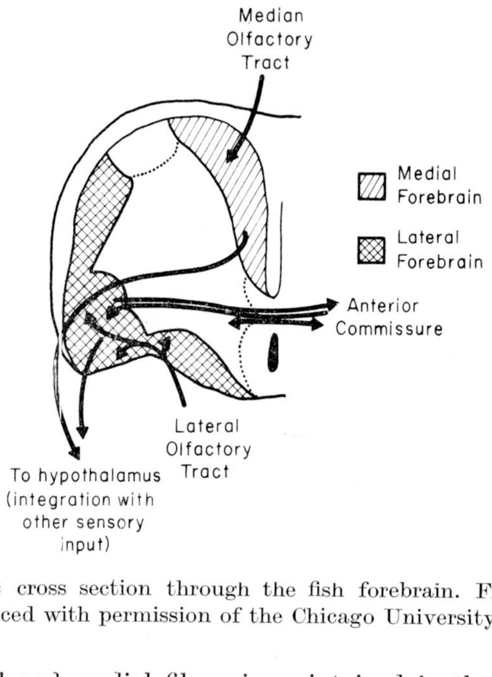

Fig. 4 Schematic cross section through the fish forebrain. From Schnitzlein (1968). (Reproduced with permission of the Chicago University Press.)

between lateral and medial fibers is maintained in their connections with the forebrain; the medial bundle goes to dorsal medial forebrain and the lateral bundle to the lateral forebrain (Fig. 4). Preserving the separation of connections might be of significance in preserving any coding occurring in the olfactory bulb.

There is no one accepted terminology to describe the various areas of the forebrain; possible homologies with higher vertebrates are still debated. Schnitzlein (1968) defined six major nuclear regions of the telencephalon—dorsolateral, dorsal, dorsomedial, ventromedial, ventral, and a central area.

Both the role of the forebrain as an olfactory center and the neural development of the forebrain vary greatly among fishes. The well developed olfactory system of cartilaginous and bony fishes is reflected in the degree of elaboration of the anterior portion of the forebrain in the area where olfactory tract fibers enter. However, many bony fishes show another direction of forebrain development—namely elaboration of the caudal dorsolateral area. This development can occur without there being significant increase in connections with the olfactory system; at the same time, however, caudal forebrain enlargement is accompanied by a great increase in thalamic and hypothalamic connections (Aronson, 1963). Such an enlarged posterior forebrain seems to serve as a higher order integrating center for non-chemosensory receptor systems. Behavioral studies of forebrain-ablated animals support this speculation (Bernstein, 1970). Furthermore, Bruckmoser (1973) mentions electrophysiological evidence of non-olfactory afferent connections to the forebrain.

In sharks olfaction and the lateral line are the primary sensory systems. However, they show very little neural interaction with one another. In teleosts the primary connections from olfactory areas of the forebrain are with the thalamic and hypothalamic areas providing the neural network for olfactory-gustatory interaction (Herrick, 1905). This network further connects with efferent branchial motor nerves facilitating effective feeding behavior (Ariens-Kappers, *et al.*, 1936).

B. Taste

1. *Taste Buds and Their Distribution*

Vertebrate taste buds show much basic similarity; in fishes clusters of a dozen to twenty cells in the epidermis are gathered into a structure that resembles a closed flower bud on a stalk of connective tissue; the latter leads the taste nerve through the dermis. Some of these cells are taste cells usually with one large microvillus at their apex, others are supporting cells with shorter, more numerous apical protuberances. They are underlain by cuboid, oval, or flattened basal cells. Recent electronmicroscopy (Graziadei and de'Han, in press) suggests that frog taste buds may have both afferent and efferent connections and thus, that their function could be centrally modulated. When one cuts a taste nerve the taste buds degenerate; when the nerve regenerates taste buds form again. The dynamics of taste cell induction require elucidation even though their natural turnover has been studied (Beidler, 1963).

Some fishes have taste buds all over their bodies with even those on their tails being innervated by a branch from a cranial nerve. Consequently one may find among fishes very peculiar ramifications of cranial nerves, especially in the *Ramus recurrens* of the facial nerve discussed in detail by Freihofer (1963). Taste buds innervated by this and other specially modified cranial nerves and located on the fish's body surface,

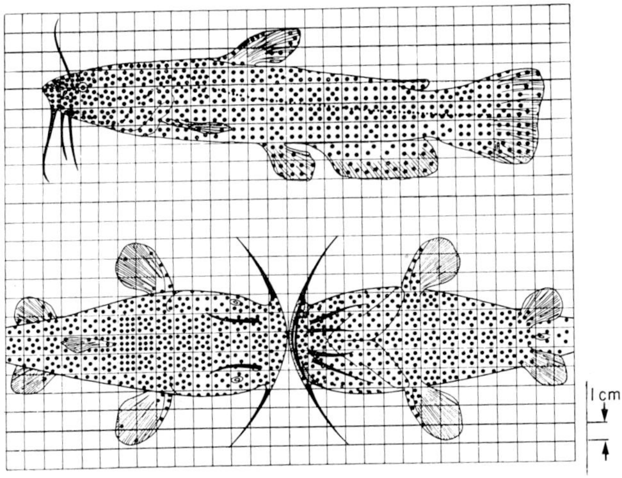

Fig. 5(a)

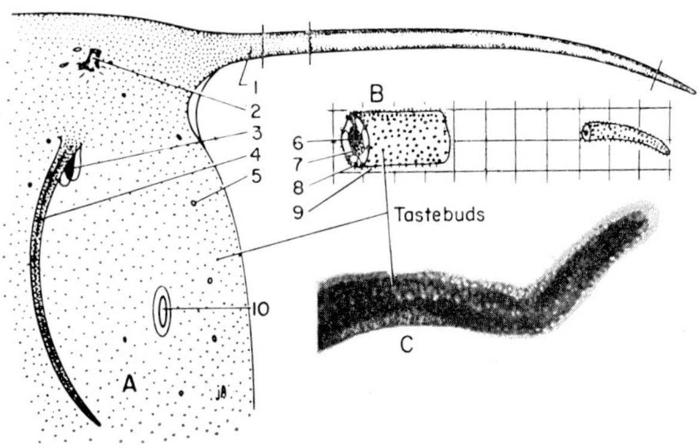

Fig. 5(b)

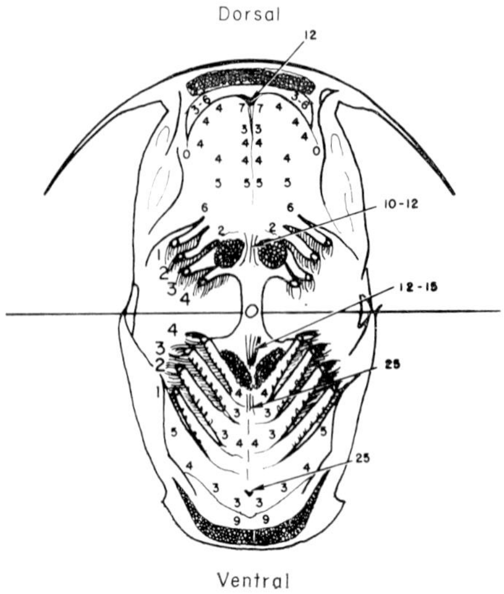

Fig. 5(c)

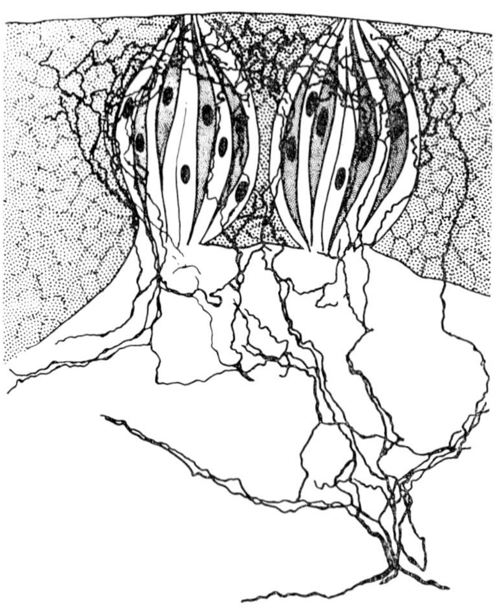

Fig. 5(d)

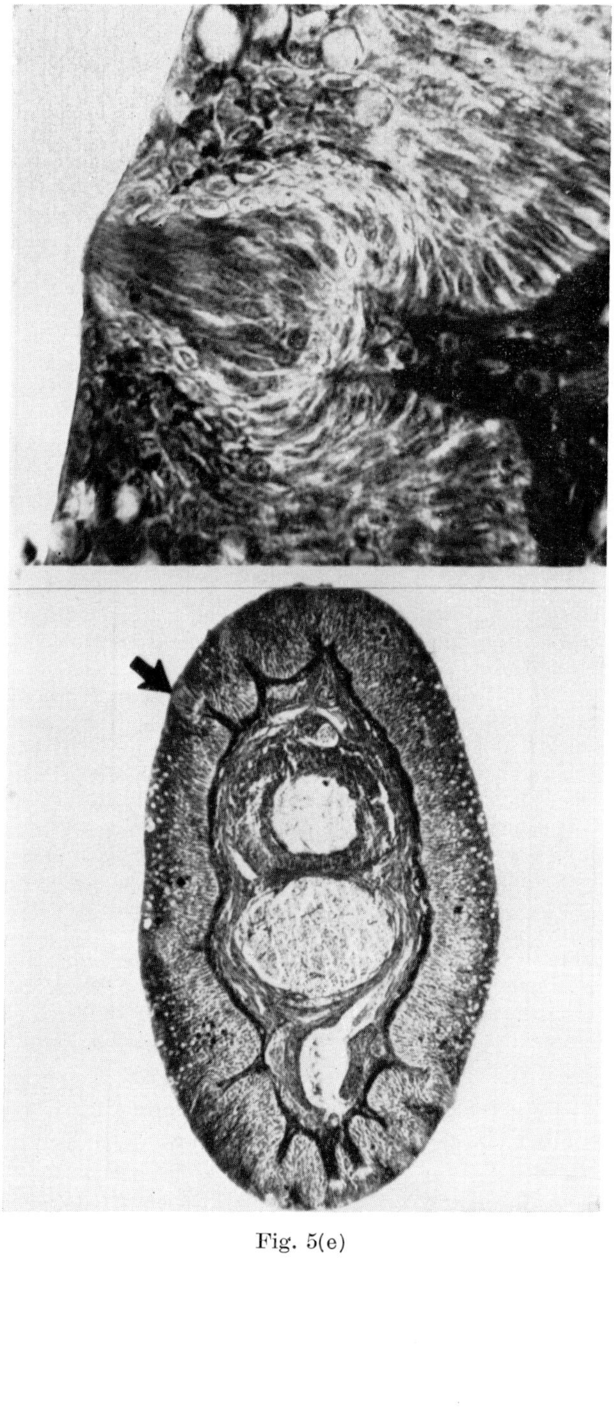

Fig. 5(e)

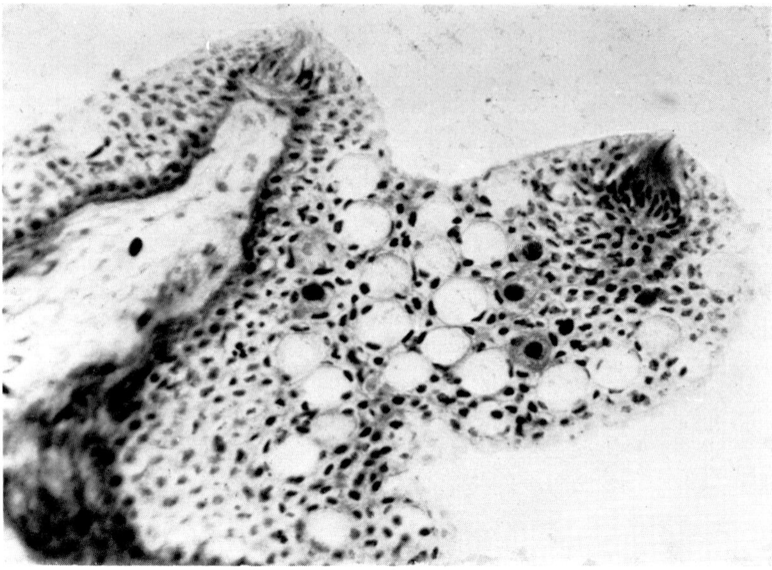

Fig. 5(f)

Fig. 5 Distribution and anatomy of fish taste buds. (a) Distribution of taste buds on the skin of an adult catfish, *Ictalurus natalis*. Scale of Grid = 1 cm. Each dot represents 100 taste buds. (Number of dots per square = number of taste buds per mm^2.) Solid black = more than 15 taste buds per mm^2. (b) Distribution of taste buds on head, lip and maxillary barbel (from Atema, 1969). A. 1:1 representation of taste buds (linear enlargement 5 ×); B. enlargement of the parts of the maxillary barbel (10 ×); C. photograph of the distal end of the maxillary barbel (15 ×); 1. maxillary barbel; 2. inflow naris; 3. outflow naris; 4. nasal barbel; 5. pore of lateral line canal; 6. supporting cartilage; 7. branch of cranial nerves V and VII; 8. taste bud; 9. mucus with protruding taste hairs; 10. eye. (c) Distribution of the taste buds in the mouth and pharynx. Numbers indicate taste buds per mm^2; for that area. High concentrations on ridges are shown by arrows and attached numbers. (d) Innervation of taste buds and surrounding skin in the barbel of a Sturgeon from Lagler *et al.* (1962). (Reproduced with permission of John Wiley and Sons.) (e) Top: Taste bud on barbel of Bullhead; Bottom: Cross section through barbel of Bullhead. (f) Elevated taste bud on lingual papilla of a Searobin. (e) and (f) authors' photo.

fins or other appendages such as barbels or oral cirri, may greatly exceed in number those that are located in the mouth and gill cavities. For instance, in a 22 cm yellow bullhead (*Ictalurus natalis*) we found about 175,000 taste buds on the barbels and the body as opposed to 20,000 in the mouth and pharynx (Atema, 1969). The codfish family shows a similar prevalence of body taste buds, compared to oral ones.

Aside from their distribution, taste buds may occur in clusters, in patches, or singly. They may be elevated on an oral papilla (Fig. 5) or

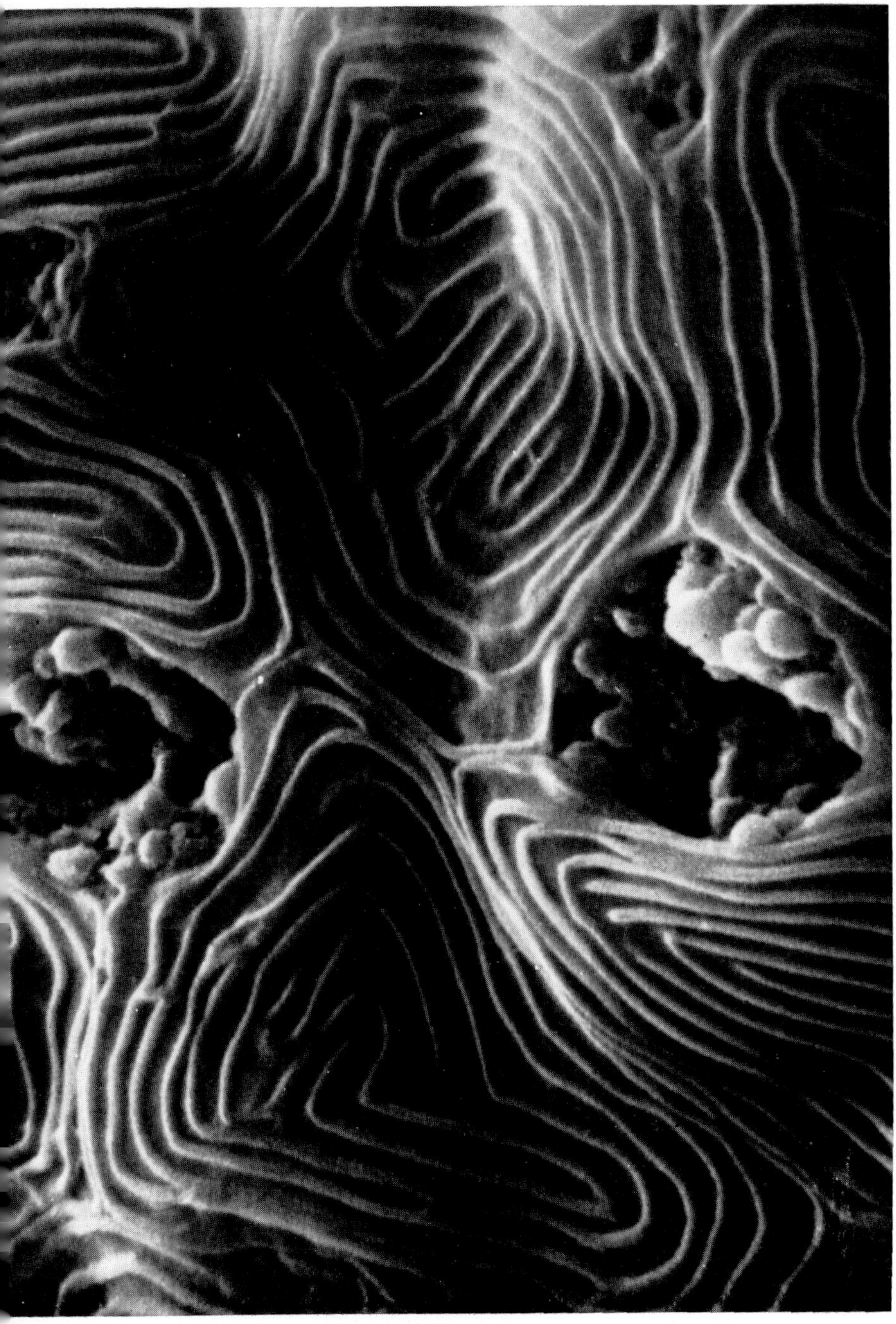

Fig. 6 Taste buds between scales of the Guppy; two round holes are observed interrupting the scale pattern and each represents the pore of a taste bud. The tiny bulb-like structures seen inside these holes correspond to the tips of microvilli which characterize the free portion of a taste cell. Courtesy P. Graziadei (magnification 9600 ×).

sunken in pits, even in openings among scales (Fig. 6). Consequently, their sensory apices may emerge slightly above the surface of the skin (Fig. 5) or be in a depression. We doubt that these varying anatomical appearances reflect functional or physiological differences. However, the occurrence of taste buds on appendages and on the body of fishes—a unique feature among vertebrates—enables the species so endowed to use taste for navigation and as a distance sense, that is to detect and locate food (see below) as well as to decide whether or not to ingest and then swallow it.

2. *Taste Nerves and Taste Centers in the Central Nervous System*

a. *Spinally innervated taste sensors.* The searobins and their relatives (*Triglidae*) have peculiarly modified pectoral fins (Fig. 7). These have limited but pronounced taste capabilities (Scharrer, *et al.*, 1947; Bardach and Case, 1965). Taste stimuli may be perceived by the abundant free nerve endings such as can also be found around taste buds (Fig. 5) but it is more likely that they are mediated by special sensory cells on the fin rays. The latter seems likely because Whitear (1965) described small methylene blue staining cells in the scaleless skin of many fishes, including searobins. The cells have certain properties, a microvillus, peculiar granules, lobate nuclei, that implicate a taste function for them (Whitear, a and b 1971); they are, however, singly too small for investigation with present electrophysiological techniques.

The tasting appendages of searobins are innervated by spinal nerves from ganglia that resemble in size certain anterior brain centers (Fig. 7); the ganglia represent a truly remarkable specialization of the vertebrate central nervous system, reminiscent of the sacral ganglion of Jurassic sauropod reptiles. Nerves emanating from them go largely tailward, in the spinal cord, an anatomical fact corroborated by behavior observations: upon very local and selective stimulation of the taste sensors on the fins, the fish begins to dig and move its tail, while the same stimulus applied selectively to the lips leads to gulping and swallowing (Bardach and Case, 1965).

b. *Cranially innervated taste sensors.* The central connections of cranial gustatory fibers are in the dorsal medulla; there is some separation of information from fibers of external and internal taste buds. Those emanating from buds in the mouth and on the lips and, if present, on the head barbels and body travel in cranial nerve VII and end more anteriorly in the dorsal medulla. Fibers from the pharnyx, gills, gill arches, and gill rakers belong to cranial nerves IX and X and terminate in the caudal medulla.

From the above two sensory centers in the medulla fibers connect to motor areas, also within the medulla. Development of these connections

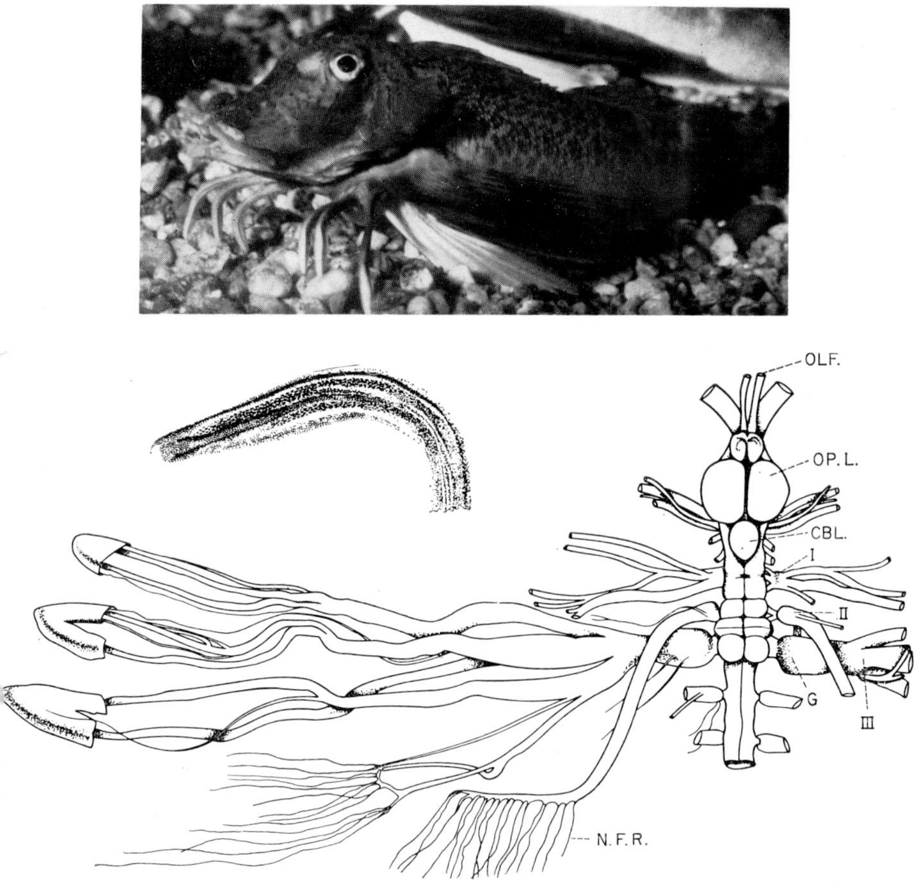

Fig. 7 The spinally innervated taste system of Searobins (Triglidae). (*Top*) The Searobin (*Prionotus carolinus*) showing its 3 anterior, moveable pectoral fin rays on which taste sensors are located. (*Bottom*) Anatomical details of Searobin brain and anterior spinal cord, after Morrill (1899). OLF = olfactory tract. OP. L = optic lobe. CBL = cerebellum. I, II, III = spinal nerves. G = large ganglion of spinal nerve III, innervating free rays. NFR = nerves of fin rays.

suggests that gustatory information may be primarily used to direct reflex actions of lips, jaws, and gills. Further connections of taste centers of the medulla go through fibers to the secondary gustatory nuclei located at the base of the cerebellum and optic tectum and thence to the hypothalamus where they allow integration of gustatory information with other sensory information—visual, olfactory, lateral line, and inner ear.

In fish which utilize gustatory information more extensively these pathways are extremely well developed; the primary ganglia in the medulla show increased size and complexity, in some cases resulting in the development of gustatory lobes. The anterior lobe is called the facial lobe receiving external information, and the caudal (or vagal) lobe receives mostly oral taste fibers.

The pattern of development of gustatory centers reflects characteristics of species behavior and the extent of use of gustatory information. In the cod, for example, vagal lobe development is very elaborate. Distinct centers receive information from the mouth (median portion of vagal lobe) and from the taste buds on the body (lateral vagal lobe). Direct connections to motor centers allow rapid precise movements (Ariens-Kappers, *et al.*, 1936). In contrast, the catfish shows elaboration of both vagal and facial lobes as well as interactions between gustatory and olfactory centers. In addition, the catfish's use of tactile information is reflected in gustatory connections with those centers.

III. Physiology of Smell and Taste

A. Thresholds and Effective Stimulants

So-called odor specialists such as the silk moth are so sensitive that one or two molecules of its pheromone suffice to elicit a reaction, but this sensitivity falls off drastically when the compound is slightly altered. Fishes, while also employing sex pheromones, show a far wider range of olfactory capabilities than insects; they are therefore considered to be odor generalists like other vertebrates. This does not imply lack of sensory acuity, however, because fishes respond to highly diluted or dispersed scents.

In a remarkable test small eels learned to associate phenylethyl alcohol, a constituent of rose oil, with the heaven-like reward to an eel of being allowed access to a dark tube. So dilute was the scent to which they reacted ($2{\cdot}8 \times 10^{-18}$ or nearly 2×10^{3} mol/cm^{3}) that there could not have been present more than two to several odorant molecules at one time in the 1–2 mm^{3} nasal capsule (Teichmann, 1959). Other volatile substances to which fishes have been trained and found to respond at low concentrations are eugenol, citral, and methanol (Teichmann, 1962).

Similar dilutions can be assumed to elicit smell triggered responses under more natural conditions: The dispersal reactions of Elritzen to "Schreckstoff" (fright substance) (von Frisch, 1941) can be elicited by a dilution of skin extract of approximately 2×10^{-11}. In view of the fact that the nasal chambers of a minnow of 5 cm length is far smaller than that of an eel of about 20 cm—the size of those employed by Teichmann—one may again assume that relatively few molecules of the substance(s) in the Schreckstoff extract are involved in the reaction.

Schools of skipjack tuna (*Katsuwonus pelamis*), a fish usually considered to feed by sight, would engage in frantic food search upon the introduction into their round tank (10 m diameter, 75,000 liter capacity) of 100 ml of colorless sea water in which one specimen of their customary food fish had been rinsed for several minutes (personal observation). The dilution can hardly be calculated but it may approach that just mentioned for the trained eels and the untrained, frightened minnows.

Taste thresholds of fishes are far higher; Glaser (1966) used serial dilution techniques and conditioning to ascertain the taste acuity of the minnow *Phoxinus* (Table 1). The fish, while endowed with extra-oral

Table 1. Comparison of taste thresholds of man (after von Skramlik, 1948) with those of the minnow *Phoxinus* (from Glaser, 1966)

Substance	*Capabilities in Mole/l*		Superiority of the Fish Over Man
	Man	Minnow	
Raffincse	—	1/245,7C0	
Sucrose	1/91	1/81,920	900-fold
Lactose	1/16	1/2560	160-fold
Glucose	1/13	1/20,480	1575-fold
Galactose	1/9	1/5120	569-fold
Fructose	1/24	1/61,440	2560-fold
Arabinose	1/13	1/15,360	1182-fold
Saccharin	1/9091	1/153,6000	169-fold
Quinine hydrochloride	1/1030928	1/24,576,000	24-fold
Sodium chloride	1/100	1/20,480	205-fold
Acetic acid	1/1250	1/204,800	164-fold

taste buds has far less of them than a bullhead or a hake (*Urophycis*). To ascertain whether summation of impulses from very numerous taste sensors would lower the threshold, we tested the quinine sensitivity of anosmic bullheads; it lay around 1×10^{-4} molar, two orders of magnitude higher than for smell. Cysteine hydrochloride, a substance not tested by Glaser (1966) but one which we knew to be a good taste stimulant was still tasted by anosmic fishes at 1×10^{-6} molar. With its sense of smell intact the fish would sense a dilution of 1×10^{-7} molar.

Certain substances, then, are both smelled and tasted by the fishes and distinction between smell and taste substances seems to be less pronounced among them than in higher vertebrates, a phenomenon that is not unexpected considering their aquatic existence (Bardach, *et al.*, 1967a). Among the chemicals that elicit both taste and smell responses cysteine is approached by naive bullheads while quinine is avoided. Quinine stimulated taste buds, but depressed spontaneous discharges in the olfactory tract. Cysteine stimulation evoked increases in the discharges of both smell, higher order and taste, secondary neurons.

The substances presently known to evoke responses of both smell and taste sensors of fishes are listed in Table 2. Even salt solutions affect

Table 2. Compounds to which smell and taste organs of fishes respond electrophysiologically with notations on behavioral reactions to them

Compound	Fish	Smelled Behaviorally	Tasted Behaviorally	Source(s)
L-Alanine	Bullhead	?	?	1
L-Cysteine	Bullhead	×	×	1, 3
L-Glutamine	Bullhead	?	?	1
L-Methionine	Bullhead	×	×	1, 3
L-Proline	Atlantic Salmon	?	× Does not initiate feeding	2
L-Serine	Bullhead	?	?	1
Sodium Chloride	Bullhead	×	×	1, 3
Potassium Chloride	Bullhead	?	?	1
Hydrochloric Acid	Bullhead	?	?	1
Acetic Acid	Bullhead	×	?	1
Quinine HCl	Bullhead	×	×	1, 3

1. Bardach, *et al.*, (1967); 2. Sutterlin and Sutterlin (1970, 1971); 3. Bardach and coworkers, unpublished.

both chemical senses, leading one to suspect that salts, in the aquatic domain, may play an important chemosensory role. Yet the behavioral significance of the sensory overlap between smell and taste and its functional correlations in the central nervous system have not been investigated.

Smell and taste overlap need not be restricted to food stimuli although this seems to be the case in the bullhead where the smell system is primarily involved in social behavior. In the Pacific salmons, amino acids some of which are likely to be smelled as well as tasted, as they are in the Atlantic salmon (Sutterlin and Sutterlin, 1970, 1971), are also implicated in smell-triggered avoidance behavior. Idler, *et al.*, (1961) tested mammalian skin rinses in which there was valine, leucine, isoleucine, cystine, and L-serine. The latter was the most potent constituent in eliciting a temporary cessation of upstream migration by representatives of all five species of *Oncorhynchus*; the reaction supposedly protects fish against predation by bears, seals and the like (Idler, *et al.*, 1961).

Another function of low molecular weight amino-compounds is to attract predatory fish to their prey. Kleerekoper and Mogensen (1959) isolated from the water that bathed two species of trout and a perch, respectively, various amines and amino acids. One of them, called "amine F" and later found to be isoleucine methyl ester (Kleerekoper, personal communication, Nov. 24, 1971), attracted the sea lamprey and marine predators alike. Thomas (1970) isolated a dipeptide of hypotaurine and glycine which she named arcamine from the mussel, *Arca zebra*. This dipeptide elicited exploratory feeding behaviour in reef fishes.

Schreckstoff, mentioned above as triggering dispersal of minnows and occurring in the club cells of Ostariophysean skin is a soluble substance or mixture of substances of unknown nature, but likely of small size (Schutz, 1956); it is smelled because severance of the olfactory nerve abolishes the reaction.

Given the broad spectrum of sensitivity of the vertebrate nose and the barely fathomed redundancy and variety of responses of its individual neurons (Lettvin and Gesteland, 1965; Nanba, *et al.*, 1966), one suspects that biologically meaningful scents—triggering feeding as well as social behavior—are only rarely single compounds but more often mixtures of several of them, each at its own level.

Food smells do, in fact, seem to represent specific mixtures of different compounds. At the aquarium in Oita (Japan) we measured the acceleration in swimming speed of captive yellowtail (*Seriola quinqueradiata*) upon presentation of food scents. The only significant speed-up, from about 1 to 5 m/sec., was observed when a rinse of the fish's customary food, a species of *Stolephorus*, was released. Rinses from other genera of fishes, of synthetic scents of molluscs, as well as presentation of several pure compounds, amino acids, and trimethylamine oxide (TMO) produced no changes in swimming speed.

Even man recognizes, by smell, different species of freshly caught fish—the reader can prove this to himself by sniffing at a fresh carp and comparing its scent to that of a pike, an eel or a trout. Fish can do the same (Göz, 1941) and they can also recognize by smell individuals of their own species (Todd, *et al.*, 1967; Todd, 1971). Individual, and probably species recognition, as suggested by the yellow-tail observation presented above, are more likely to rely on signature-tune-like mixtures of several compounds than on a particular single one. However, single compounds in mixtures may also be important as in the avoidance by salmon of serine, among other amino acids, and the amine F, prey attractant, in a welter of other related low molecular weight compounds. Combination of signature-tune and single attractant characteristics of scents would go a long way to explain the uncanny smell performances of fishes. We still need to unravel what the social and food scents might be, a fascinating exercise in its own right but also one with interest to sport and commercial fishing industries.

B. Electrophysiological Correlates of Smell and Taste

1. *Smell*

There are fewer electrophysiological studies of fish than of mammalian chemical senses even though it is technically possible to place electrodes at any level in the fish smell and taste systems. In fact, as Nanba, *et al.*, (1966) say, fishes are very suitable experimental animals to study smell. "Dosage control of stimulants and their application is easier in water than in air, and fishes have a broad spectrum of olfactory sensitivities. In addition they can learn easily to distinguish between different smells." Electrophysiological investigations of fish smell and taste are on the increase.

Electrode monitoring in the olfactory system of fishes has been concentrated on the epithelium and its sensory cells and on the fibers of the tract, for the obvious reason that these provide the easiest access and permit clearer recording than is possible from the extremely fine fibers of the nerve or the complex of cells in the bulb.

Slow potential changes in the olfactory epithelium of fishes were shown to occur on smell stimulation (Shibuya, 1960); they are somewhat faster in fishes with a pigmented olfactory epithelium (eel) than in the absence of pigments. The former also has a keener sense of smell. Upon stimulation and coincident with the slow potential phenomena (EOG), one also observes changes in spike discharges from olfactory cells (Shibuya, 1960). However, largely because of the difficulties of recording from single primary sense cells, the dynamics of olfactory cell function in all vertebrates are incompletely understood. Recent electrophysiological studies of fish olfactory epithelia deal mainly with ascertaining the range of sensitivities of selected species, especially to amino acids (Suzuki and Tucker, 1971; Sutterlin and Sutterlin, 1971).

Recordings from single units of the fish olfactory tract (Døving, 1966; Nanba, *et al.*, 1966) give information on neural events after considerable integration and elaboration of information has taken place in the bulb. Nanba, *et al.*, summarize their results from *Abramis brama* and *Carassius auratus* as follows:

1. A fiber exhibits different patterns of activity when different odors are employed (temporal coding).
2. Different fibers respond to the same odor in different ways (spatial coding).
3. The response to the stimulus depends not only on the nature of the odor but also upon its concentration, the turbulence of the water stream acting on the mucosa and the spontaneous background activity acting on the fiber. However, the effect of an odor was consistently reproducible provided the other parameters did not change.
4. Many fibers discharge in rhythmic bursts, spontaneously, or on stimulation with an odor. This rhythmic activity persists after dissection of the central end of the olfactory tract.
5. Most fibers exhibit adaptation following continuous stimulation with the same odor.
6. The position of receptors belonging to the fiber being recorded can be determined by mechanical stimulation of discrete areas of the mucosa. If other receptors are so stimulated, lateral inhibition results.

Surface and depth recordings were secured from the forebrain of sharks by Gilbert, *et al.*, 1964. They found changes in wave patterns and potentials upon perfusion of the nose with fish and crab extracts as well as with amino acids, and Kleerekoper's "amine F". Further details on physiological aspects of fish olfaction can be found in Kleerekoper, 1969, who also compares electrophysiological investigations of fish smell with those done on other vertebrates, especially amphibians and mammals.

2. *Taste*

Electrophysiological correlates of taste responses in fishes, like those of smell, are generally similar to those of other vertebrates. Yet some certain peculiarities of the fish taste sense have been shown by electrophysiological means.

1. Cranially innervated taste sensors (taste buds) show a greater spectrum of responses than spinally innervated ones (free nerve endings and/or small presumptive chemo-sensory cells). Spinal taste nerves especially do not respond to sugars (Table 3).
2. Oral taste receptors of carp respond to gaseous CO_2 and CO_2 dissolved in water (Hidaka, 1970). Thus they seem to be involved in the regulation of respiration.

3. Fish possess distinct taste sensors for dilute as well as more concentrated electrolytes (Konishi, 1967; Konishi and Hidaka, 1967).

4. Fish are suspected to taste "special substances" emanating from prey (Konishi *et al.*, 1966; Konishi and Zotterman, 1963); some of these may be glycerophospholipids, while others have been identified as nucleotide mono and diphosphates (Kiyohara *et al.*, in press). The matter is discussed in detail in Bardach and Atema (1971).

Table 3. Comparison of electrophysiological taste responses of hake and tomcod fins as opposed to those of the searobin*

	Hake or Tomcod (taste buds; cranial innervation)	Searobin (no taste buds; spinal innervation
Acetic acid	+	+
Sodium chloride	+	+
Sucrose**	+	0
Quinine hydrochloride	+	+
Cysteine	+	+
Homocysteine	?	0
Methionine	0	0
Cystein hydrochloride	+	+
Cystic acid	?	?
Glutathione	+	+
Aspatric acid	+	0
Alanine	+	+
Serine	0	0
Leucine	0	+
Glutamic acid	+	0
Tryptophane	?	?
Tryptamine	?	?
Phenylalanine	+	+
Glycine	+	0

*From Bardach, *et al.*, (1967); Bardach and Case (1965).
**Note especially absence of reactions to sugar in the spinal fibers.

5. Fine discrimination of taste seems to be enhanced by fishes possessing seven or more taste fiber types which respond to single compounds or groups of compounds (Konishi and Zotterman, 1963, their Table 1 p. 219).

The following additional characteristics of taste are probably not restricted to fish even though they were demonstrated through electrophysiological experiments with them. Different species of fishes possess different numbers of fibers that respond to basic taste substances, for instance 79 out of 114 carp taste fibers responded to sugars (Konishi and Zotterman, 1963) while we found only 5 out of 51 comparable sweet fibers in the bullhead.

Such fiber distributions are, no doubt, the genetic endowment of a species that has evolved through food preferences or restrictions. These

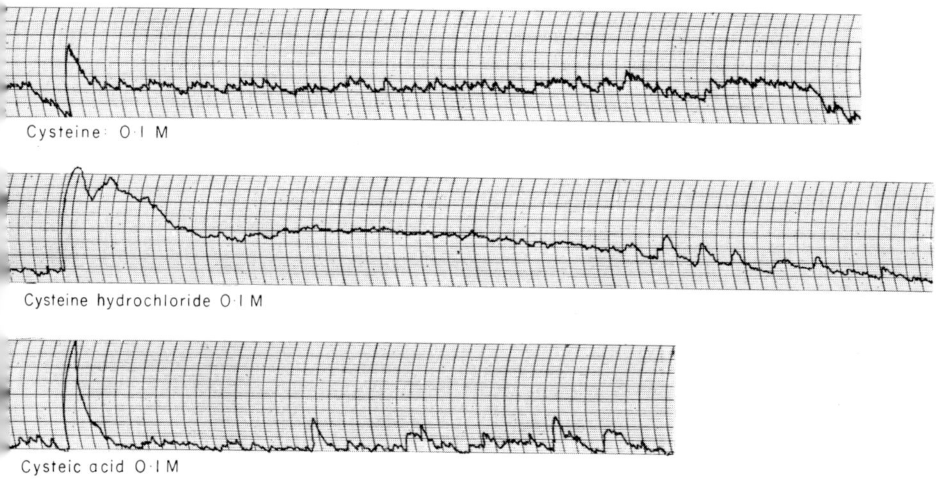

Fig. 8 Comparison of integrated responses of several bullhead fibers to cysteine and related compounds. See text for explanation. Time constant—0·02 sec; Recorder speed—5 mm/sec equals one chart division. From Fujiya and Bardach (1966). (Reproduced with permission of the Japanese Society of Scientific Fisheries.)

adaptations are comparable to olfactorily triggered innate food preferences of neonate snakes (Burghardt, 1970) even though they are not as striking. They probably also exist in the smell domain of fishes.

Anatomical, behavioral, as well as electrophysiological observations point to taste as being in many fish species a more discriminatory sensory modality than in other classes of vertebrates. We offer here, as additional evidence for this tenet the integrated recordings of responses from bullhead barbel taste fibers to three relatively similar substances (Fig. 8).

All records from several taste fibers of fishes also show mechanoreceptor responses and close associations exist between taste and touch centers in the fish's medulla. Some taste fibers are also sensitive to peripheral heating or cooling. Taste appears to be relatively more important in fishes than it is in the remaining chordates.

IV. Behavior

A. Feeding Behavior

All fish use the chemical senses in feeding, albeit to a varying extent. In highly visual fishes, these senses only test the food immediately prior to ingestion. Whatever the senses involved, feeding behavior shows a stereotyped sequence of behavioral components. Wunder (1927) defined three stages: (1) alert, (2) locate, and (3) control; Atema's (1971) similar classification expanded the control component, to wit: (1) arousal (alert), (2) orientation and search (locate), (3) food uptake into the mouth, and (4) food ingestion. Chemical sensing can mediate any or all of these steps, partially or completely.

1. *Arousal*

Arousal to the presence of food is the necessary first step in the feeding sequence. A variety of behavioral responses can be indicative of the animals having become alert to the presence of food; among them changes in respiratory patterns, swimming patterns, activity levels or barely perceptible tensing of the body or fins. Arousal by a chemical component of the food has been demonstrated in a number of studies. Hobson (1963) noted that dogfish sharks (*Musculatus canis*) only approached bait from downstream, indicating that a chemical stimulus rather than visual or auditory one is involved, since the latter would extend equally in any direction. Tester (1963) demonstrated changes in activity of sharks in response to water extracts from damaged and undamaged prey as well as to blood and some other chemical compounds.

Parker and Sheldon (1913) demonstrated that cutting the olfactory tract or blocking the nares of dogfish prevent their response to bags of food which previously excited them. They also noted that the catfish (*Ameirus* = *Ictalurus*) does not respond to food when its olfactory tract is cut, but with a normal olfactory system it shows the ability to notice and locate a food packet even with an impaired gustatory system (removed barbels). Moray eels (*Gymnothorax*) are unable to detect or locate food when their nares are plugged (Bardach, *et al.*, 1959). Killifishes (*Fundulus heteroclitus*) are also unable to recognize food when their nares are stitched shut (Parker and Sheldon, 1913).

Kleerekoper and Mogensen (1963) noted that the characteristic increase in the activity of lampreys following the introduction of food extract (or water which had contained live specimens of the lampreys' normal prey) does not occur if the olfactory apparatus is blocked. Likewise, the searobin and the hake require an intact olfactory system to become alert and begin a search (Bardach and Case, 1965). Some response of the gustatory fins was seen in response to a food extract in anosmic animals but no search ensued.

Pipping (1926, 1927) examined the response of fish to food odors in relation to the morphology of their olfactory organ. Animals with no other method but forward progression to irrigate the olfactory epithelium show less response to food extracts than those with a highly developed nasal apparatus, such as an elongated nasal chamber and pronounced anterior and caudal openings, as well as accessory sacs or cilia to move the stimulus over the olfactory epithelium.

Although most work indicates that olfaction mediates arousal, some fishes are capable of recognising the presence of food without their olfactory apparatus, presumably through their gustatory system. Bardach, *et al.*, (1967b) and Atema (1971) demonstrated that the yellow bullhead (*Ictalurus natalis*) shows essentially normal feeding behavior with its olfactory epithelium cauterized and even feeds essentially normally with its forebrain removed. With the experimental removal of the forebrain there is some increase in latency of arousal, possibly attributable to the general role of the forebrain in arousal, as opposed to an olfactory role of the forebrain (Atema, 1969).

Leopard sharks, *Triakis semifasciata*, with both nares blocked by paraffin, still show increased locomotor activity and zest for food upon the introduction of whole squid (Carney, personal communication). Normal sharks were present during the observations. Therefore the normal fish might have aroused the deolfacted ones; yet the activity of normal sharks under other conditions do not trigger activity in the test fish.

2. *Food Search*

The mode of food location shows a wider diversity of sensory dependence than the other steps of feeding behavior. In a significant number of fishes, the primarily visual ones, there is usually little or no involvement of chemosensory information in the localization of foods. For example, the killifish which Parker and Sheldon (1913) described show such emphasis on visual stimuli that they will chase water droplets down the side of an aquarium.

Fishes with lesser vision depend upon their chemical senses to locate food, some upon olfaction and some on gustation. Parker (1914) observed the searching pattern of dogfish which had one naris blocked. The handicap led the fish to circle continually to the side of the intact organ, instead of swimming its normal figure eight pattern. Interestingly the fish required no longer to locate the food source than an unencumbered subject.

Gustation seems to play a primary role in the orientation and location of food for a number of fishes, which show taste buds over the body and fins (*Ictalurus* and other catfishes, some minnows, blind cavefishes, hake, tomcod, and other gadoid fishes). Schemmel (1967) in discussing the adaptation of *Astyanax* to cave dwelling, noted that the blind cave dweller had developed a more extensive gustatory innervation of the jaw and ventral region than his river dwelling cousin. He suggested that this more extensive gustatory sense serves to better locate, rather than

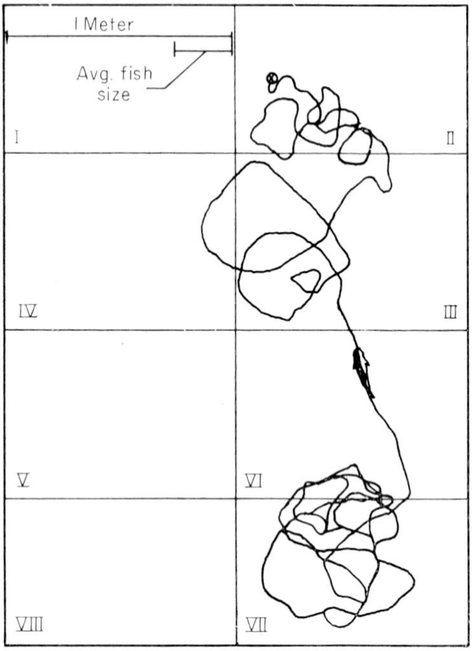

Fig. 9 Photographed swimming path of a bullhead unilaterally deprived of taste buds (white portion of fish in sketch). Method of obtaining record is illustrated in Fig. 10 and described in Bardach, *et al.*, (1967b). (Reproduced with permission of the American Association for the Advancement of Science.)

to detect their food. In the bullhead, there is no behavioral evidence for the necessity of olfactory information during search. In experiments conducted in tanks of up to 5 m in length, there seems to be little role for olfaction in any aspect of the bullhead's feeding behavior. The marine goatfish (*Parupeneus porphyreus*), endowed with taste buds on its chin-barbels, will also sense food and begin searching for it when its nares are cauterized or its olfactory nerves are cut (Holland, personal observation). Unilateral destruction of gustatory receptors, however, results in circling similar to that seen in the dogfish after occlusion of one nostril (Fig. 9) (Bardach, *et al.*, 1967b).

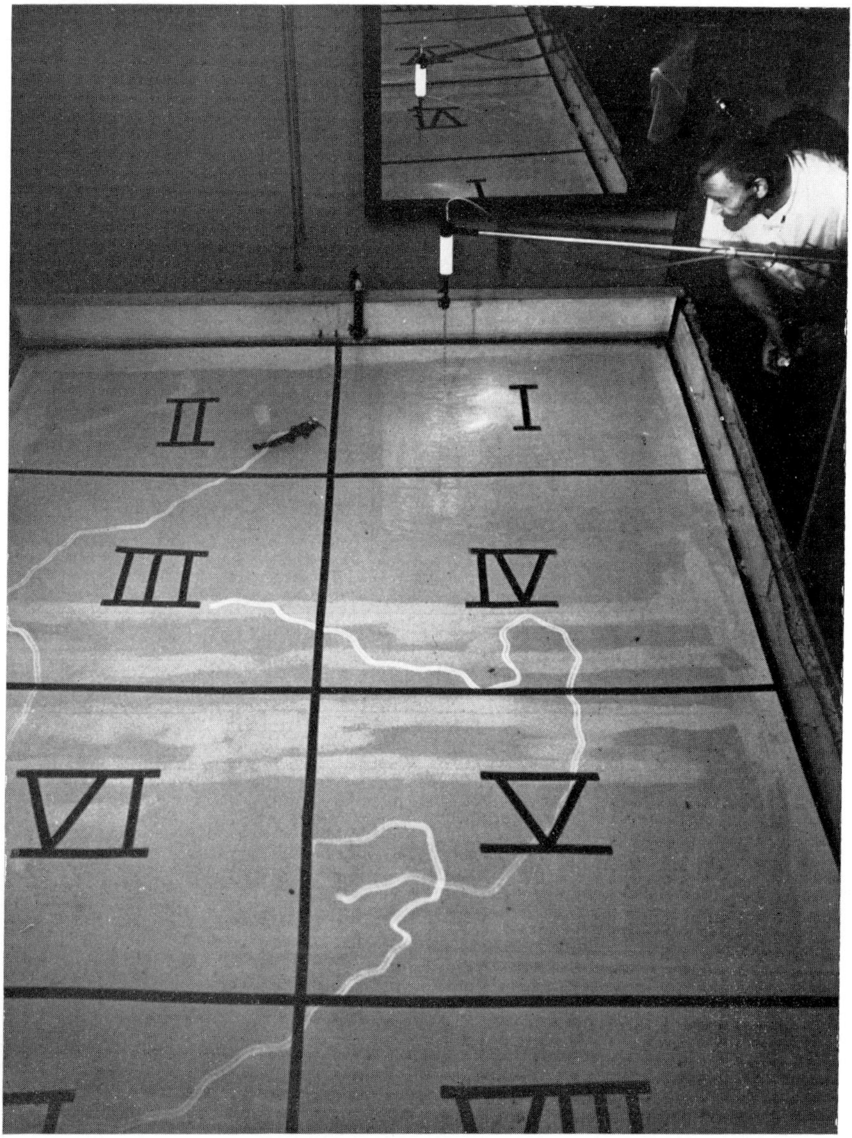

Fig. 10 Swimming path of a blinded bullhead with fluorescent plastic platelet affixed on head, searching for chemical stimulus release point and photographed in ultraviolet light; stimulus has fluorescin in it so as to show its distribution. Method is described in Bardach, *et al.*, (1967b). (Reproduced with permission of the American Association for the Advancement of Science.)

3. *Orientation*

Nares as well as taste sensors are bilaterally symmetrical; the latter, as in the case of the bullhead, where they are most concentrated on the tips of the barbels, could permit true tropotaxis to occur (comparison of chemical concentration between two points). Similarly, the distribution of taste buds over the entire body, from snout to tail would make it possible for a fish to detect a fore-aft difference in stimulus concentration (Fig. 10). We noted that bullheads find the source of a chemical faster and with less deviation from a straight path in still, as opposed to flowing, turbulent water (Fig. 11). It appears therefore that certain fishes, under certain conditions rely on a tropotactic mode of *gustatory* searching.

Olfactory tropotaxis has been questioned by Kleerekoper (1969) but is suggested by Mathewson and Hodgson (1971). They obesrved that nurse sharks (*Gynglymostoma cirratum*) react to chemical stimulation by swimming directly toward the source of a mixture of 0·1 molar glycine and 0·1 molar trimethylamineoxide. "Their reactions begin with an extra gill beat, followed by a brief closure of gill openings as they start toward stimuli. The pectoral fins are occasionally used to orient the initial lunge toward a stimulant by larger nurse sharks. Observations suggest that they can, indeed, discriminate differences in concentrations impinging upon sense organs on the two sides of the head, and that this ability can be used to 'home-in' on a source chemical stimuli at least 40 ft away (a minimal estimate-maximum determination will have to await further studies). We have not yet done the appropriate experiments, removing barbels or anesthetizing olfactory organs, to enable us to draw any conclusions about the receptors mediating these responses."

It is, in fact, difficult to conceive how a fish could detect absolute concentration differences in the streams of water through its two nares (except for the extreme case of the hammerhead shark with its widely separated nares). Rather, a mechanism might operate whereby the fish through reflexes attempts to equalize the sensory inputs it receives through both nares. This would effectively orient the fish toward the source of the odor. Temporary enhancement of discrepancies of input from right and left nares respectively would assist in the discrimination necessary to achieve stimulus equalization. Two such mechanisms for facilitating olfactory tropotaxis in fishes have been noted: the first involves inhibition of one olfactory bulb while the nares connected to the opposite one are stimulated, and vice versa (Hara and Gorbman, 1967). The second is the differential flow controlled by the epithelium itself. Dr. Chen in our laboratory demonstrated that stimulation with a food substance of one nasal chamber of a bullhead slows down the flow

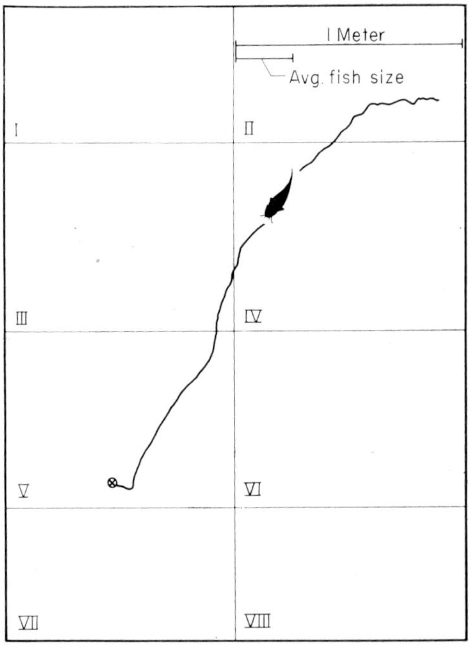

(a)

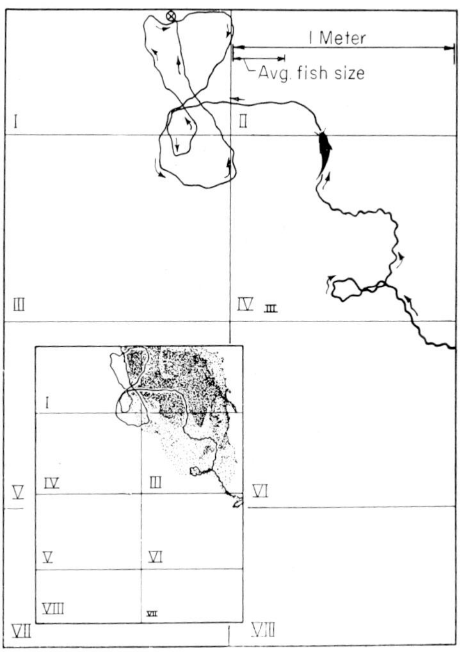

(b)

Fig. 11 (a) Swimming path of a bullhead toward release point of a taste stimulant in still water where substance will spread concentrically. Note almost straight path. (b) Swimming path of a bullhead toward a release point of a taste stimulus in flowing water; insert shows uneven stimulus distribution. Note that the path is more complex than in (a).

of water through the opposite chamber (Fig. 12). The difference in flow is caused by differences in the ciliary beat in the two chambers. Transection of the anterior commissure or unilateral severing of a tract abolishes the reaction which must therefore be mediated by efferent fibers.

Another explanation of the manner in which fishes orient to food scents might involve phobotaxis, the tendency to change the angle of

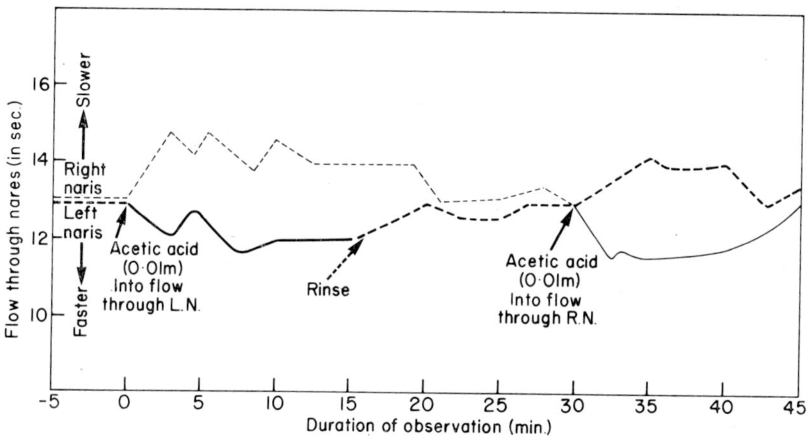

Fig. 12(b)

Fig. 12 (a) Method of obtaining time record of water flow through nares in fishes. Top insert shows micropipette directed at incurrent nares; appearance of a nonstimulant dye at posterior nares as monitored with binocular microscope determines flow-through time. (b) Record of unilateral flow control in bullhead nares upon stimulation with acetic acid achieved through differential ciliary action.

deviation in response to changes in stimulus concentration. Kleerekoper (1967a, b) demonstrated, with *Scyliorhinus* and *Mustelus diplodus*, that the angle of deflection in their swimming paths decreases with the presence of natural food odors. Teichmann (1959) noted a clear phobotactic response in the blind eel (*Anguilla anguilla*) to an odor conditioned to be associated with food.

Rheotaxes is yet another pattern of response possibly employed in food location. Lemon sharks (*Negaprion brevirostris*) for instance do not appear to "home in" on the source of chemicals (again, a mixture of 0·1 molar glycine and 0·1 molar trimethylamineoxide used in comparative behavior trials with nurse sharks, above (Mathewson and Hodgson, 1971) but rely on chemical stimulation to trigger rheotaxis. Upon olfactory arousal the shark heads "upstream" instead of following a circular path in which only one half of the circle brings the animal to swim against the prevailing current. They are slower in locating the source of the stimulus than the nurse shark which appears to be capable of discriminating right-left differences in chemical stimulus concentration.

The presence of a current not only facilitates the delivery of the stimulus but it can apparently provide a directional cue also. Kleerekoper (1967a) showed that a 10% flow difference between two compartments allows a fish to localize an odor source in the compartment from which it emanated. Odor alone isn't sufficient and no rheotaxic behavior is noted in the absence of a food odor. It is noteworthy in this context that the blind cavefish (*Amblyopsis*) shows positive rheotaxis to currents of 2–7 cm/sec (Poulson, 1963). The fish may rely entirely on the lateral line in its food search or it may employ both nose and lateral line combined in the manner suggested above. In the food-poor environment of caves this would be a time and energy saving procedure.

4. *Food Uptake and Ingestion*

Testing of food palatability relies on gustation (Bardach and Atema, 1971) as is suggested when moray eels reject quinine treated anchovies (Bardach, *et al.*, 1959), sharks reject deodorized squid (Tester, 1963), and yellowtail immediately spit out denatured food pellets (Bardach, personal observation).

Taste cells and mechanoreceptors lie close to one another on lips and appendages, as well as in the mouth, and their centers in the central nervous system also allow close comparison between what food feels and tastes like. Ablation studies emphasized that touch and taste qualities are important and that food uptake and ingestion proceeds in a sequence of reflexes.

Ablation studies also support the concept of separation of function between the facial and vagal gustatory lobes. In bullheads, ablation of the facial lobe blocks food uptake into the mouth only, while vagal ablation blocks food ingestion (Atema, 1971). The muscles involved are not impaired since other jaw action, such as yawning, appears normal; also impairment is specific to the complex behavior trains of either food uptake or ingestion. Atema interpreted these results as evidence that external taste buds mediate food uptake while internal gustation mediates ingestion. A fish with its facial lobes intact and its vagal one impaired will pick up food morsels and immediately spit them out again, only to pick them up again; one fish persisted in doing so for over an hour (Atema, 1971).

B. Reproductive Behavior

Chemosensory information plays a role in a number of aspects of fish behavior; in parental protective and mating, in homing to spawning sites, and in various nonreproductive social behaviors, including individual recognition and territorial behavior, species recognition and schooling, as well as in the recognition of predators and of stress and danger signals.

1. *Homing*

The role of olfaction in the homing of fishes has been extensively studied. The best known instance of homing is that of migratory salmon, which will travel across an ocean in their life cycle. Many other fish species also show cyclic return to a specific site; for instance wild brown trout, suckers, charr, sculpin, and rainbow trout. Other nonmigratory fish will demonstrate homing if displaced from their normal locale; among these are longear sunfish, certain gobies, green sunfish, largemouth bass and white bass (Kleerekoper, 1969).

Harden Jones (1968) differentiated between the maintenance of a home position and the return to a home area either after being displaced, or during cyclical migration. He noted that the role of the senses in the maintenance of a home position has not been carefully studied, while much recent work deals with the problem of the homing response. Thorpe (1956) proposed vision to be of major importance in the maintenance of a location in the home area. Exceptions might be *Protopterus* and the blind cave fish (Thines, 1955). Hasler (1966) also included in his hypothesis of salmon homing, a role for vision, not only in the final recognition of the spawning site but also in recognition of points along the upstream migration of the salmon and in orientation during the ocean phase of travel. He hypothesized that the evening and night rest spots at points along the downstream movement of the fingerlings, serve to visually imprint upon them the landmarks at which they rest. In their later return home they use the imprinted signposts for orientation.

But the most extensively supported theories hold that olfactory information is necessary for upstream migration. The question of whether a stream could in fact be recognized by its chemical components was tested by Hasler and Wisby (1951). They trained blind minnows to discriminate between the waters from two forks of a stream. Not only were the fish capable of performing this task, but the qualities of the stream remained such, over time, that the discrimination could still be made. They also noted that similar training was successful with salmon fingerlings.

Other behavioral data also suggest chemical cues in the homing response. Migrating salmon could be removed from a stream and tested for their response to their homestream and other streams. Idler, *et al.*, (1961), and Fagerlund, *et al.*, (1963), demonstrated that these salmon increased their swimming speed and dispersed their shoals upon introduction of their homestream water, but not upon introduction of an equal amount of stream water from another spawning site.

The nature of the chemical homing stimuli was examined. They were found to be volatile, dialyzable, heat labile compounds, destroyed by active charcoal or by storage (Idler, *et al.*, 1961; Hasler and Wisby, 1951; Creutzberg, 1961).

The olfactory sense is the major receptor of this chemical input. Hasler not only noted that the ability of minnows to discriminate between streams depends upon an intact olfactory apparatus but he also cites various field studies on returning salmon which indicate that blocked or cauterized nares, or severed olfactory tracts greatly impair the ability of the damaged fish to correctly reach their spawning grounds (Hasler and Wisby, 1951; Wisby and Hasler, 1954; Kleerekoper, 1969). Imprinting of coho salmon smolt with morpholine has been achieved. The volatile compound does not occur in nature but is well perceived by the fish's olfactory sense. The imprinted animals returned to the training site using geographical and olfactory orientation mechanisms (Cooper and Hasler, 1974).

A different approach to the study of the role of olfaction in salmon homing was taken by a group of researchers at the University of Washington in Seattle. By making electroencephalic recordings from the olfactory bulbs of salmon, caught during their upstream migration, they were able to obtain a physical measure of the salmon's response to various stream waters. They demonstrated increased bulb activity in response to stimulation by waters from the spawning site but not from other streams (Hara, *et al.*, 1965; Ueda, *et al.*, 1967) (Fig. 13).

Further studies using this technique have attacked the problem of how homestream odor is remembered. Various, as yet inconclusive, studies have attempted to block the memory for homestream waters or change the fishes' original preference to that of another spawning site (Oshima, *et al.*, 1969; Gorbman, personal communication, 1971).

In order to explain how salmon swim ceaselessly many hundred miles upstream an olfactory triggered rheotactic response has been suggested (Kleerekoper, 1969). The homestream odor triggers it and if the fish makes an improper turn into the wrong stream its drive to swim upstream ceases with the absence of the "scent" of its spawning site. It would then be carried back downstream to a place where it will again encounter its homestream odor and begin its ascent anew. No role in homing has been suggested for gustatory senses.

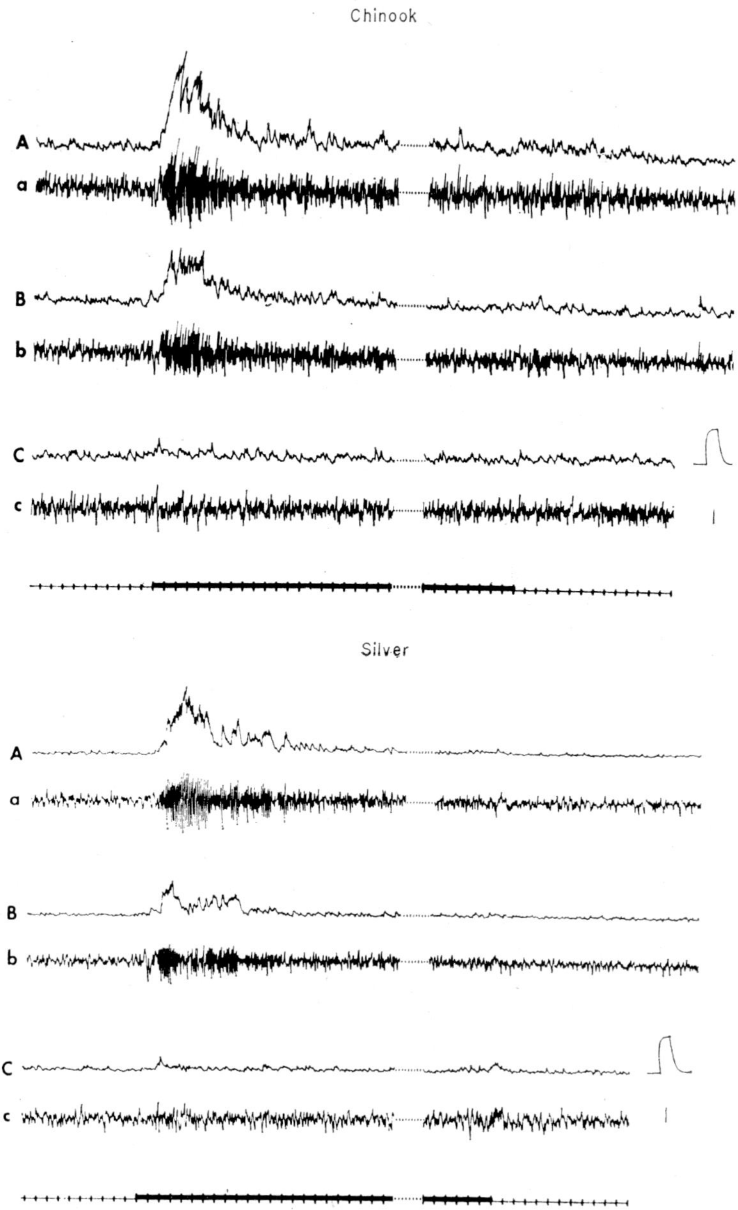

Fig. 13 Electroencephalographic response of the olfactory bulb to home water. EEG readings were taken from the olfactory bulb of adult migratory salmon shortly after they had arrived at their spawning sites. The lower tracings (indicated by a, b, and c) are the actual EEG readings; the upper (indicated by A, B, and C) are integrated tracings of those EEG's. The top two (A and a) are in response the salmon's home stream; the next two (B and b) are in response to waters nearby the spawning site; and the last two are responses to water from a source unrelated to the home stream (from Ueda, *et al.*, 1967). (Reproduced with pergmission of Peramon Press.)

2. *Mating*

Mating pheromones have been reported in certain fish. In one species of gobies, *Bathygobius soporator*, small quantities of seawater holding a gravid female are sufficient to elicit courtship displays by the male (Tavolga, 1956). In the gurami (*Colisa calla*), a similar phenomenon was demonstrated. Even though visually isolated, water from a female fish triggered the bubblenest blowing in the male which is preliminary to courtship (Mainardi and Rossi, 1970). In the shad also, a substance from the female has been implied in triggering courtship behavior in the male (Roule, 1926).

Similarly, chemicals are sometimes used in the discrimination between sexes. Seitz (1940) noted that the male *Haplochromis multicolor* mouths the urogenital region of other fish of the same species, and suggested that such behavior might assist in the discrimination between sexes. MacGinitie (1939) demonstrated the role of chemicals in such discrimination with the blind goby, *Typhlogobius californiensis*. A pair of these fish will reside together in burrows created by a shrimp. They will remain relatively tranquil in their burrow, but if another blind goby enters, a vicious fight ensues. If the intruder is a male the resident male attacks; if a female enters, the female attacks. The following simple experiment showed that this recognition was based on chemical cues. When a perforated cellophane bag containing either a male or a female goby was placed in the burrow, the information in the water alone was sufficient to inform the resident pair of the sex of the intruder. The appropriate resident would then attack the bag.

Other research has demonstrated that chemicals attract the female to the male in the lamprey (Roule, 1926), in a blenny (*Blennius pavo* Eggert, 1931), and in a glandulocaudine fish (Nelson, 1964). In the salmon the opposite is believed to occur; the female attracts the male (Belding, 1934). Strong circumstantial evidence for female sex pheromones that attract the males also comes from certain deep sea fishes (many ceratioid anglerfishes and *Cyclothone microdon*). Their males must attach themselves to the larger females before reaching maturity; aside from various other attributes of sexual dimorphism, the olfactory organs of the two sexes differ markedly, as described by Marshall (1971): "After metamorphosis, the males become rapidly macrosmatic. When fully developed each olfactory organ contains a stack of broad sensory lamellae, which are open to the sea through two wide nostrils. The olfactory organs of female ceratioids and *Cyclothone* are reduced to minute proportions. Matching this extreme dimorphism, the olfactory nerves, bulbs and forebrain are strongly developed in males but regressed in females." Finally, it has been suggested that pheromones from one male stimulate courtship in other males (Leiner, 1930; Losey, 1969). The conclusions have been questioned, however (Bardach and Todd, 1970).

The olfactory rather than the gustatory sense is suspected of handling all chemical information in courting, primarily because of the low levels of chemicals involved. Tavolga (1956) demonstrated that deolfacted gobies no longer respond to the female odor. The sensory system involved has not been critically investigated in the remainder of the mating pheromones studies.

In some cases, however, gustation might play a role. As mentioned above, Seitz (1940) observed mouthing by *Haplochromis multicolor* in sex discrimination, an activity that could involve gustatory rather than olfactory cues. Gouramis (*Colisa*) use their long filamentous pelvic fins to stroke each other during their courtship (Forselius, 1957). These fins, although primarily tactile, also bear taste buds (Scharrer, *et al.*, 1947; Steinbach, 1950). This might suggest some role for gustation in their mating.

3. *Parental Behavior*

The role of chemical senses, appears to be significant in the care of the young in some species of fish. Especially cichlids recognize their own young, pay them ample attention, and in some cases frequently mouth them.

Kühme (1963) demonstrated chemical recognition of young jewel fish by their parents, which will select a water source that contains their young over other water outlets (Fig. 14).

Myrberg (1966) showed a similar behavior in *Cichlasoma nigrofasciatum*, a highly visual fish. He maintained that later parental recognition is a visual learning phenomenon which depends upon earlier, instinctive chemosensory recognition. MacGinitie (1939) noted that parent blind gobies will not eat their young while other gobies will, suggesting some specific recognition based on nonvisual—probably chemical—cues. Künzer (1964) also demonstrated recognition of their own young by a dwarf cichlid.

It is unclear whether these parent fish smell or taste their young; olfaction would be suspected but the frequent mouthings of the young by many cichlids has given rise to the suggestion that gustatory information may play a role (Myrberg, 1966). Certainly gustatory information would be expected to allow discrimination by the parent between the young and food; perhaps it is also used to discriminate between their own and another brood.

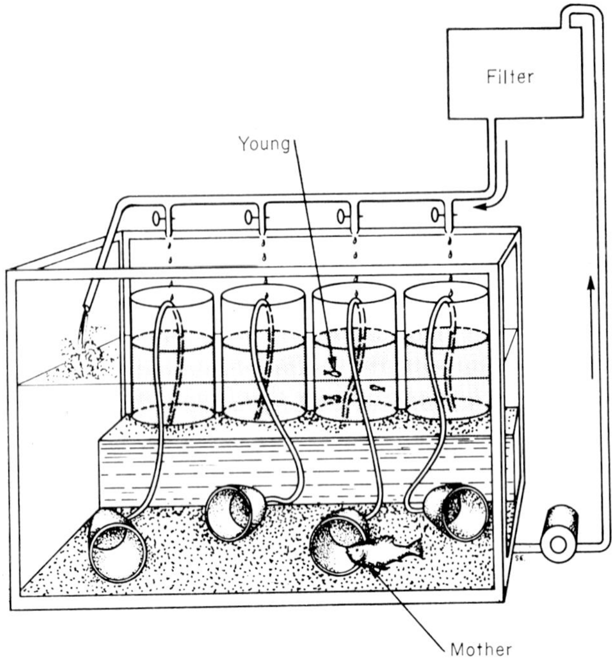

Fig. 14 Experimental arrangement for testing pheromones from young cichlids; after Kühme (1963).

C. Social Behavior

1. *Species Recognition*

The significance of chemical cues in the recognition of one fish by another has been reported only recently (Todd, *et al.*, 1967). Chemicals can serve for recognition of conspecifics as in the phenomenon of schooling, and for the recognition of other species. Fright pheromones (Schreckstoff) exist which are utilized in signaling the need for escape. Furthermore fish can recognize individuals within their own species, which implies that they possess more elaborate social behavior than one normally gives them credit for.

a. *Species recognition and schooling.* The degree to which individuals within a species will aggregate covers a wide spectrum. Some fish are hermits who remain alone and will not tolerate others of their species while others show highly uniform movements; they move in schools which appear like the movements of one, not of many fish. Groupings of fish can be elicited by external factors, such as food or light, or they may be based upon mutual attraction (Allee, 1931). In a very loose sense, the term schooling includes any grouping of fish which is based upon mutual attraction.

The most significant sensory system in schooling is the visual one. Impairment of vision dissolves schools even in fishes which otherwise show little use of vision. A partial list of fishes which have been demonstrated to depend upon vision to school is found in Atz (1953). Other senses, however, are also involved in schooling, albeit to a far lesser extent. *Menidia menidia* and *Alosa pseudoharengus* will still swim within a school for a few seconds when blinded, although the response is not maintained much longer (Shaw, 1970). Though not thoroughly investigated, it has been postulated that schools may be kept coherent by tactile, chemical, auditory, and lateral lines cues; some of the species listed by Atz (1953) in this regard are: *Ameiurus* = *Ictalurus*, *Plotosus*, *Tilipia*, *Pneumatophorus*, and *Rutilus*.

While not having a pronounced schooling response, Wrede (1932) demonstrated that *Phoxinus* will prefer to enter one compartment in a three compartment maze which had previously held a species mate. Hemmings (1966) revealed a similar preference among cichlids, and Kühme (1964) demonstrated some chemical stimuli in the schooling of young jewelfish. Keenleyside (1955) showed a similar phenomenon for the rudd, *Scardinius erythrophthalmus*, and ascertained that it is specific to members of the same species. He then deolfacted the experimental fish and the preference disappeared. These studies suggest that olfaction can provide an initial fish-to-fish attraction and maintain individuals in proximity to one another.

In some fishes schooling can be maintained in the absence of olfaction. Bullheads will still show the same aggregating-schooling behavior with their olfactory tracts cut. Forebrain ablation which affects the central connections for the olfactory system, in contrast to deolfaction which just eliminates the nose, also seems to have some effect on schooling behavior (Aronson, 1963). The effect is not so much a decline of, but rather a change in aggregation, suggesting a regulatory, modifying role for the forebrain such as intimated by Aronson and Kaplan (1968). After forebrain ablation schooling behavior is more reflexlike and stereotyped than it was before (Berwein, 1941; Jansen, 1933; Shaw, 1970).

A possible role for taste in fish aggregations might be related to the detection of external conditions. Increases in CO_2 content disperses aggregations in cichlids (Baerands and Baerands van Roon, 1950) and variations in pH of the water can increase or decrease their aggregations (*Hemichromis*) (Noble and Curtis, 1939).

b. *Species recognition and fright reaction.* Related to schooling and aggregation is the effect of chemical signals which result in dispersal. The fright reaction was first noted by von Frisch (1941) in the minnow *Phoxinus* which shows strong schooling behavior. The presence of an injured minnow would cause a "fright reaction" in the rest of the school, a wash of the extract of pieces of such an injured minnow would effectively scatter a group of minnows that had gathered to feed. The school would disperse, gather together at a distance from the source of the alarming odor, and flee.

The nature of the alarm reaction was clearly defined. The fact that an extract was effective eliminated visual, tactile, or lateral line cues. In a series of tests it was demonstrated that skin, particularly the epidermis, contained the alarm substance (Schreckstoff).

Alarm pheromones and reactions are widespread throughout the Ostariophysian fishes, including the Cypriniformes and also Siluriformes that inhabit the marine as well as fresh-water environments. In some of the most recently evolved characins the alarm pheromones are absent, although they are Ostariophysi; it has been suggested that the ability to produce the pheromone has been secondarily lost in this group (Pfeiffer, 1963). Within the order Cypriniformes, the alarm pheromone from one species will elicit an alarm reaction in other species. The strongest reactions occur between closely related species, even when they are widely separated geographically. The interfamilial responses are only slight (Schutz, 1956), which suggests that the chemical structure of the pheromone probably varies from species to species, being least similar in the most widely separated groups.

Steven (1959) reported the presence of the alarm pheromone and reaction in a non-Ostariophysian fish, the atherinid, *Hepsetia stipes* (*Atherinomorus stipes*). Skinner, *et al.*, (1962), described an alarm reaction in another atherinid, the top smelt, *Atherinops affinis*. Rosenblatt and Losey (1967) failed to find the pheromone, or reaction, in the top smelt and suggested that Skinner, *et al.*, described an alarm reaction caused by experimentally induced stress, possibly resulting from mechanical disturbances.

The source of the alarm pheromone in the Ostariophysian fish appears to be similar throughout the group. Superficial injury of the skin shows that the epidermis contains the alarm substance (von Frisch, 1941). There is a particular type of club cell in the epidermis which is unique to those species which produce alarm pheromones. According to Pfeiffer (1960) these club cells only release their contents to the body surface upon injury. The barbel epidermis of the carp and some catfish contains no club cells or only a few very small ones, while the body epidermis of these species is abundantly supplied with these cells. The fright reaction is elicited by body skin, but not by barbel skin (Pfeiffer, 1963). Very little work has been done in ascertaining the chemistry of alarm pheromones. Boiling the substance for 5 minutes does not affect its potency, but boiling continued for 10 minutes reduces its effectiveness to about one-fifth of its normal level (von Frisch, 1941). It has been suggested that the alarm substance is a purine or pterin-like substance (Hüttel, cited by Pfeiffer, 1963). These substances are concentrated only in xanthophores which are most numerous on the dorsal surface. Since ventral skin is an equally effective source of the alarm pheromone, Hüttel's suggestion is most likely incorrect (Pfeiffer, 1963). Tucker and Suzuki (1972) analyzed olfactory responses of catfish to the alarm substance. Their electrophysiological and biochemical evidence suggests that the alarm substance of any one species is a mixture of several compounds present in species specific ratios; some of these compounds may be amino acids and oligopeptides. Other volatile ones remain undetermined.

The alarm substance for *Phoxinus* is not volatile although extremely soluble in water (Schutz, 1956). Despite this fact the pheromone is detected by the fish's sense of smell. When the olfactory nerve is severed, *Phoxinus* is not able to detect its presence.

Von Frisch (1941) and Kulzer (1954) demonstrated that the reaction is olfactory, not gustatory. Deolfaction of minnows or tadpoles effectively blocks the reaction.

The effect of the fright substance on schooling was discussed by Thines and Vandenbussche (1966) who noted that there is a diurnal variability in schooling intensity. This variability is reflected in a corresponding variation in the sensitivity of the school to alarm substances.

2. *Individual Recognition*

The role of olfaction for individual recognition within species has been studied by Göz (1941). He demonstrated that a member of the species *Phoxinus* was capable of discriminating, after training, between individuals of its own species.

The possible significance of individual recognition in social behavior was first noted by Todd, *et al.*, (1967). They chose a fish which shows a highly complex social pattern, and well developed chemosensory system, the yellow bullhead. The bullhead, like *Phoxinus*, shows the ability to discriminate between individuals of his species. Todd demonstrated that the olfactory system is necessary for this recognition to occur. The substance which allows recognition was first thought to be either in the slime or in the urine. But slime alone could be used to condition fish to recognize one another. Thus, the skin contains the chemicals necessary for individual recognition just as it contains the source of Schreckstoff (Bardach and Todd, 1970). In fact the observations of Tucker and Suzuki (1972) on the species specific nature of the alarm substance point to a functional explanation of individual odors. These are likely to be genetically determined mixtures of various compounds in the slime, the concentration of which is specific to the individual with perhaps one of the other chemicals being absent altogether. This hypothesis of how one individual becomes chemically different from another is much more plausible than having individual recognition rely on only one specific compound for each animal.

Individual recognition could be affected by experience. It has been observed that a stable community could be upset by submitting one member to some external stress. For example, if the dominant fish of a three-fish community is removed and placed in a tank with another larger fish, and consequently severely mauled in several fights, his status is changed upon return to his own community. In contrast, a fish who is just removed and later returned does not show a status change. Similarly, testing the effect of such stress on the recognition of one by another fish indicates that a trained fish no longer recognizes a certain other individual after it is subjected to a losing battle.

The question remains unresolved as to whether the skin odor changes as a result of a losing battle, or whether some stress odor is superimposed which counter-balances a previously positive odor response. Evidence that these responses are olfactory was demonstrated by deolfaction of the fish, which subsequently were unable to recognize known tank mates.

In communities of fish, certain territorial boundaries can be observed. If one member of a community is removed, the introduction of his odor alone is sufficient to elicit many of the same behavior traits in the other tank mates, as if the individual himself were present. For example, a very dominant fish would cause three other tank mates to respond by hiding in their shelters. When it was removed, and later water from its tank poured into the community tank, almost immediately the residents responded by fleeing and hiding.

Todd observed that four deolfacted fish, when placed in a large community tank took an excessively long time to establish some rank order. They fought tenaciously among themselves as if they had difficulty learning to recognize each other. He noted that when they finally began to establish a community there was some regeneration of the sensory cells in the nares (Todd, 1968).

Later, observations were made of some established communities in which the members were deolfacted after some residence time together. In these cases no clear disorientation of the tank mates resulted, suggesting that perhaps previously learned patterns did not require olfactory information for maintenance of adequate communal behavior.

Observations were made by Atema (1969) of the effect of deolfaction, forebrain removal, and degustation upon the interaction among bullheads. His results indicate that taste plays no role in social behaviour among bullheads. They are known to explore one another with their highly gustatory barbels, but this is probably a tactile response.

Both deolfaction and forebrain removal show effects upon aggressive interactions. Deolfaction enhances aggression in many cases as does forebrain removal. However, it is also noted that these same operations do not cause this effect consistently. The results are variable enough to suggest that other natural factors, as yet unexplored, strongly affect aggression and communal interaction among the fish.

V. Future Research on Fish Chemical Senses

A. Fish as Subjects for Basic Smell and Taste Research

Details of smell and taste stimulus transduction processes are unknown. Fish lend themselves well for the electrophysiological probing of all the components of their smell and taste systems including the appropriate portions of their brains. If the brain correlates of fish smell and taste senses were to become the areas of concern for a number of researchers interested in nervous and sensory system evolution, one should proceed to devise simple and reliable stereotactic instruments for the fish brain.

Furthermore, the relatively long olfactory nerve of certain fishes, e.g., the gar (*Lepisosteus*), invites investigations of the basic conduction characteristics of fine unmyelinated fibres in the vertebrate nervous system. Also, the olfactory epithelium of fishes is more accessible than that of land vertebrates.

The peculiarly wide array of fish forebrain development, from being essentially a series of smell ganglia in some and a set of elaborate integration centers in other species invites work on questions of comparative brain development, even to throw light on evolution of the highest primate brain centers.

Taste is generally considered a simpler sense than smell, at least its gamut is far more restricted. Basic taste processes are, therefore, somewhat better understood and workable models exist for some of them (Beidler, 1954). Attempts have been made to extract from taste buds of mammals certain substances instrumental in the binding of a taste stimulus to its sensory substrate and, therefore, in eliciting the initial biophysical event with which sensation begins (Dastoli and Price, 1966; Price, 1969). Fishes with barbels and well developed external taste buds (most *Ictalurid* catfishes) would be a far better experimental animal for taste biochemistry and biophysics than mammals. Their thin barbels carry thousands of taste buds and they therefore have a far greater proportion of sensory in relation to neutral skin tissue than the mammalian tongue. The large amount of nonsensory skin on the tongue was a serious obstacle in, and led to criticisms of, the just mentioned extraction experiments.

Spinally innervated taste sensors in high concentrations on the free fin rays of searobins (Triglidae) make them suitable for electrophysiological research. Considerable interest exists in this variant of taste sensors in contrast to taste buds proper, as the investigation of their capabilities may be relevant to an understanding of congenital taste anomalies in man in which normal taste bud development is lacking.

B. Ecological Implications of Research with Fish Smell and Taste Capabilities

Emphasis should be placed on determining the nature of effective stimuli that elicit feeding behavior and certain social behavior traits of fishes. Certain scents that trigger feeding seem to contain one or more low molecular weight substances. Presently available, especially chromatographic techniques of chemical analysis would make possible the detailed description of compounds and mixtures that lead to feeding responses, with resulting practical applications to fish management.

Similarly, as the slime of fishes is widely involved in social behavior we should try to characterize the chemical nature of the compounds in slime by detailed analysis of the mucus, including the histochemistry of the skin cells most likely involved in the production of these substances.

Some research interest might also lie in the apparent greater overlap between smell and taste reactions in fishes as compared to terrestrial vertebrates. Beyond categorizing further which materials are both smelled and tasted one would wish to gain insight into the central coordinating mechanisms of these two related sensory modalities. The fish nervous system can serve well to elucidate them, certainly with the use of a fish stereotactic instrument, already mentioned. Here again, understanding of how smell and taste work together in food search, food finding, and ingestion is bound to have practical applications to fish husbandry.

As one surveys the species of fish on which experiments and observations that deal with their chemical senses were made, one notes the relatively small number of them and the fact that far more of them reside in fresh than in salt water. Neither fact is surprising; first, most biologists live in or near population centers, inland—hence the emphasis on freshwater forms. Secondly, marine fish are often more delicate, and, thirdly, once a certain animal has shown some interesting attributes or proved itself to be hardy and suited to laboratory life it is this animal which will be employed in preference to others in many aspects of research. In the piscine realm the minnow *Phoxinus* (Elritze) and the goldfish are good examples of these restrictive characteristics.

However, the living resources of the sea are increasingly important in maintaining and improving the quality of human nutrition. At the same time we are also polluting the seas to an ever greater extent. Some chronic pollution effects are believed to work through interference with chemical senses of aquatic animals, either by impairing sensory performance or as mimics of natural stimuli (Blumer, 1970). More emphasis must be placed on finding out what role the chemical senses play in the lives of many of the important marine fishes. For instance, what are the taste and smell capabilities of commercially important herbivores, planktonivores, and piscivores. Such knowledge, together with other more basic information on their chemical senses (e.g., the role of olfaction in arousal to food, and more generally, the role of the olfactory forebrain) would help to eliminate one serious bottleneck in mariculture, the economically sound feeding of many fish species.

Another critical area in which the chemical senses are known to play a role is that of mating and parental care. Less information exists on it than on feeding behavior although it is likely to be as critical for performing sound mariculture with many species, as is detailed knowledge of the role of smell and taste in feeding.

References

Allee, W. C. (1931). "Animal Aggregations, a Study in General Sociology". University of Chicago Press, Chicago.

Arien-Kappers, C. V., Huber, C. and Crosby, E. (1936). "The Comparative Anatomy of the Nervous System of Vertebrates Including Man". Macmillan, New York.

Aronson, L. R. (1963). The central nervous system of sharks and bony fishes with special reference to sensory and integrative mechanisms. *In* "Sharks and Survival" (P. W. Gilbert, ed.), pp. 165–241. D. C. Heath and Co., Boston.

Aronson, L. R. and Kaplan, H. (1968). Functions of the teleostean forebrain. *In* "The Central Nervous System and Fish Behavior" (D. Ingle, ed.), pp. 107–125. University of Chicago Press, Chicago.

Atema, J. (1969). The chemical senses in feeding and social behavior of the catfish, *Ictalurus natalis*. Ph.D. Dissertation, University of Michigan.

Atema, J. (1971). Structures and functions of the sense of taste in the catfish (*Ictalurus natalis*). *Brain, Behavior and Evolution*, 273–294.

Atema, J., Todd, J. H. and Bardach, J. E. (1969). Olfaction and behavioral sophistication in fish. *In* "Proceedings of the Third International Symposium on Olfaction and Taste" (C. Pfaffman, ed.), pp. 241–251. Rockefeller University Press, New York.

Atz, J. W. (1953). "Orientation in Schooling Fishes". Presented at a conference on "Orientation in Animals" to the American Museum of Natural History, Washington, D.C., February 6–7.

Baerends, G. P. and Baerends van Roon, J. M. (1950). An introduction to the study of the ethology of cichlid fishes. *Behaviour Suppl.* **1**.

Bardach, J. E. and Atema, J. (1971). The sense of taste in fishes. *In* "Handbook of Sensory Physiology" (L. Beidler, ed.), Vol. IV, pp. 293–336. Springer Verlag, Heidelberg and New York.

Bardach, J. E. and Case, J. (1965). Sensory capabilities of the modified fins of squirrel hake (*Urophycis chuss*) and searobins (*Prionotus carolinus* and *P. evolans*). *Copeia* **1965** (**2**), 194–206.

Bardach, J. E. and Todd, J. H. (1970). Chemical communication in fish. *In* "Advances in Chemoreception. Vol. I—Communication by Chemical Signals" (J. W. Johnson, D. G. Moulton and A. Turk, eds.), pp. 205–240. Appleton-Century-Crofts, New York.

Bardach, J. E., Winn, H. E. and Menzel, D. W. (1959). The role of the senses in the feeding of the nocturnal reef predators, *Gymnothorax moringa* and *G. vicinus*. *Copeia* **1959** (**2**), 133–139.

Bardach, J. E., Fujiya, M. and Holl, A. (1967a). Investigations of external chemoreceptors of fishes. *In* "Proceedings of the Second International Symposium on Olfaction and Taste" (T. Hayashi, ed.), pp. 647–665. Pergamon Press, Oxford and New York.

Bardach, J. E., Todd, J. and Crickmer, R. (1967b). Orientation by taste in fish of the genus *Ictalurus*. *Science, N.Y.* **155**, (3767), 1276–1278.

Beidler, L. M. (1954). A theory of taste stimulation. *J. gen. Physiol.* **38**, 133–139.

Beidler, L. M. (1963). Dynamics of taste cells. *In* "Proceedings of the First International Symposium on Olfaction and Taste" (Y. Zotterman, ed.), pp. 133–148. Macmillan, New York.

Belding, D. L. (1934). The spawning habits of the Atlantic salmon. *Trans. Am. Fish. Soc.* **64**, 1–211.

Bernstein, J. J. (1970). Anatomy and physiology of the central nervous system. *In* "Fish Physiology" (W. S. Hoar and D. J. Randall, eds.), Vol. 4, pp. 2–90. Academic Press, New York.

Berwein, M. (1941). Beobachtugen und Versuche über das gesellige Leben von Elritzen. *Z. vergl. Physiol.* **28**, 402–420.

Blumer, M. (1970). Oil contamination and the living resources of the sea. F.A.O. Technical Conference on Marine Pollution and its Effects on Living Resources and Fishing, Rome, December 9–18.

Bruckmoser, P. (1973). Beziehungen zwischen Struktur und Funktion in der Evolution des Telencephalon. *Verh. d. Deutschen Zool. Ges.*, **66.** Jahresvers., 219–229 (Gustav Fischer verl.).

Buddenbrock, W. von (1952). "Vergleichende Physiologie. Bd I—Sinnesphysiologie", pp. 465–466. Birkhauser, Basel.

Burghardt, G. H. (1970). Chemical perception in reptiles. *In* "Advances in Chemoreception. Vol. I—Communication by Chemical Signals" (J. W. Johnson, D. G. Moulton and A. Turk, eds.), pp. 241–308. Appleton-Century-Crofts, New York.

Burne, R. H. (1909). The anatomy of the olfactory organ of teleostean fishes. *Proc. zool. Soc. Lond.* **1909**, 610–663.

Cooper, J. C. and Hasler, A. D. (1974). Evidence for Retention of Olfactory Cues in Homing Coho Salmon, *Science, N.Y.* **183**, 336–338.

Creutzberg, F. (1961). On the orientation of migrating elvers (*Anguilla vulgaris* Turk.) in tidal areas. *Neth. J. Sea Res.* **1**, 257–338.

Dastoli, F. R. and Price, S. (1966). Sweet-sensitive protein from bovine taste buds: isolation and assay. *Science, N.Y.* **154**, 905–907.

Døving, K. B. (1966). Efferent influence upon the activity of single neurons in the olfactory bulb of the burbot. *J. Neurophysiol.* **29**, 675–683.

Eggert, B. (1931). Die Geschlechtsorgane der Gobiiformes und Blenniformes. *Z. wiss. Zool.* **193**, 249–588.

Fagerlund, U. H. M., McBride, J. R., Smith, M. and Tomlinson, N. (1963). Olfactory perception in migrating salmon. III. Stimulants for adult sockeye salmon (*Oncorhynchus nerka*) in home stream waters. *J. Fish. Res. Bd Can.* **20**, 1457–1463.

Forselius, S. (1957). Studies of anabantid fishes. *Zool. Bidr. Upps.* **32**, 208–241.

Freihofer, W. C. (1963). Patterns of the Ramus lateralis accessorius and their systematic significance in teleostean fishes. *Stanford ichthyol. Bull.* **8** (**2**), 81–189.

Frisch, K. von (1941). Über einen Schreckstoff der Fischhaut und seine biologische Bedeutung. *Z. vergl. Physiol.* **29**, 46–145.

Fujiya, M. and Bardach, J. E. (1966). A comparison between the external taste sense of marine and freshwater fishes. *Bull. Jap. Soc. scient. Fish.* **32** (**1**), 45–56.

Gilbert, P. W., Hodgson, E. S. and Mathewson, R. (1964). Electroencephalograms of sharks, *Science, N.Y.* **145**, 949–951.

Glaser, D. (1966). Untersuchungen über die absoluten Geschmacksschwellen von Fischen. *Z. vergl. Physiol.* **49**, 492–500.

Gooding, R. (1963). The olfactory organ of the skipjack, *Katsuwonus pelamis. F.A.O. Fish. Rep.* **6** (**3**), 1621–1631.

Göz, H. (1941). Uber den Art und Individualgeruch bei Fischen. *Z. vergl. Physiol.* **29**, 1–45.

Graziadei, P. P. C. and de'Han, R. S. (In press). Ultrastructural morphology of frogs' taste buds. *Acta anat.*

Hara, T. J., Ueda, K. and Gorbman, A. (1965). Electroencephalographic studies of homing salmon. *Science, N.Y.* **149**, 884–885.

Hara, T. J. and Gorbman, A. (1967). Electrophysiological studies of the olfactory system of the goldfish, *Carassius auratus* L.—I. Modification of the electrical activity of the olfactory bulb by other central nervous structures. *Comp. Biochem. Physiol.* **21**, 185–200.

Harden, Jones, F. R. (1968). "Fish Migration". St Martin's Press, New York.

Hasler, A. D. (1966). "Underwater Guideposts". University of Wisconsin Press, Madison.

Hasler, A. D. and Wisby, J. (1951). Discrimination of stream odors by fishes and its relation to parent stream behavior. *Am. Nat.* **85**, 223–238.

Hemmings, C. C. (1966). Olfaction and vision in fish schooling. *J. exp. Biol.* **45**, 449–464.

Herrick, C. J. (1905). The central gustatory paths in the brains of bony fishes. *J. comp. Neurol.* **15**, 375–456.

Hidaka, I. (1970). The effect of carbon dioxide on carp palatal chemoreceptors. *Bull. Jap. Soc. scient. Fish.* **36** (**10**), 1034–1039.

Hobson, E. S. (1963). Feeding behavior in three species of sharks. *Pacif. Sci.* **17**, 145–170.

Holl, A. (1965). Vergleichende morphologische und histologische Untersuchungen an Geruchsorgan der Knochenfische. *Z. Morph. Ökol. Tiere* **54**, 707–782.

Idler, D. R., McBride, J. R., Jonas, R. E. E. and Tomlinson, N. (1961). Olfactory perception in migrating salmon. II. Studies on a laboratory bio-assay for homestream water and mammalian repellant. *Can. J. Biochem. Physiol.* **39**, 1575–1584.

Jansen, W. (1933). Untersuchungen über Grosshirnfunktionen des Goldfisches (*Carassius auratus*). *Zool. Jb. Abt. Allg. Zool. Physiol. der Tiere* **52**, 591–628.

Katsuki, Y., Hashimoto, T. and Kendall, T. I. (1971a). The chemoreception in the lateral-line organs of teleosts. *Jap. J. Physiol.* **21**, 99–118.

Katsuki, Y., Hashimoto, T., Onoda, N. and Yanagisawa, K. (1971b). Chemoreception of lateral-line organs in aquatic animals. Abstract in *Proc. internatn. Union physiol. Sci.*, Vol. **IX**.

Keenleyside, M. H. A. (1955). Some aspects of the schooling behaviour of fish. *Behaviour* **8**, 183–248.

Kiyohara, S., Hidaka, I. and T. Tamura. (in press). The Gustatory Response in the Puffer, *Fugu pardalis*—II. Single fiber analyses. *Bull. Jap. Soc. Scient. Fish.*

Kleerekoper, H. (1967a). Some effects of olfactory stimulation on locomotor patterns in fish. *In* "Proceedings of the Second International Symposium on Olfaction and Taste" (T. Hayashi, ed.), pp. 625–645. Pergamon Press, Oxford and New York.

Kleerekoper, H. (1967b). Some aspects of olfaction in fishes with special reference to orientation. *Am. Zool.* **7**, 385–395.

Kleerekoper, H. (1969). "Olfaction in Fishes". Indiana University Press, Bloomington, Indiana.

Kleerekoper, H. (1971). Personal communication.

Kleerekoper, H. and Mogensen, J. A. (1959). The chemical composition of scent of freshwater fish with special reference to amines and amino acids. *Z. vergl. Physiol.* **49**, 492–500.

Kleerekoper, H. and Mogensen, J. A. (1963). Role of olfaction in the orientation of *Petromyzon marinus*. I. Response to a single amine in prey's body odour. *Physiol. Zool.* **36**, 347–360.

Kleerekoper, H. and van Erkel, G. A. (1960). The olfactory apparatus of *Petromyzon marinus* L. *Can. J. Zool.* **38**, 209–223.

Konishi, J. (1967). Studies on the stimulation of chemoreceptors of freshwater fish by dilute solutions of electrolytes. *In* "Proceedings of the Second International Symposium on Olfaction and Taste" (T. Hayashi, ed.), pp. 667–692. Pergamon Press, Oxford and New York.

Konishi, J. and Hidaka, I. (1967). Stimulation of the chemoreceptors of the sea catfish by dilute electrolyte solutions. *Jap. J. Physiol.* **17**, 726–737.

Konishi, J. and Zotterman, Y. (1963). Taste functions in fish. *In* "Proceedings of the First International Symposium on Olfaction and Taste" (Y. Zotterman, ed.), pp. 215–233. Macmillan, New York.

Konishi, J., Uchida, M. and Mori, Y. (1966). Gustatory fibers in the sea catfish. *Jap. J. Physiol.* **16**, 194–202.

Kühme, W. (1963). Chemisch ausgelöste Brutpflege und Schwarmverhalten bei *Hemichromis bimaculatus* (Pisces). *Z. Tierpsychol.* **20**, 688–704.

Kühme, W. (1964). Eine chemisch ausgelöste Schwarmreaktion bei jungen Cichliden (Pisces). *Naturwissenschaften* **51**, 120–121.

Kulzer, E. (1954). Untersuchungen über die Schreckreaktion der Erdkrötenkaulquappen (*Bufo bufo* L.). *Z. vergl. Physiol.* **36**, 443–463.

Künzer, P. (1964). Weitere Versuche zur Auslösung der Nachfolgereaktion bei Jungfischen von *Nannacara anomala* (Cichlidae). *Naturwissenschaften* **51**, 419–420.

Lagler, K. F., Bardach, J. E. and Miller, R. R. (1962). "Ichthyology". John Wiley and Sons, New York.

Leiner, M. (1930). Fortsetzung der ökologischen Studien an *Gasterosteus aculeatus*. *Z. Morph. Ökol. Tiere* **16**, 499–522.

Lettvin, J. Y. and Gesteland, R. C. (1965). Speculations on smell. *Cold Spring Harb. Symp. quant. Biol.* **XXX**, 217–225.

Losey, G. (1969). Sexual pheromone in some fishes of the genus *Hypsoblennius* Gill. *Science, N.Y.* **163**, 181–183.

MacGinitie, G. E. (1939). The natural history of the blind goby *Typhlogobius californiensis* (Steindachner). *Am. Midl. Nat.* **21**, 489–505.

Madison, D. M., Scholz, A. T., Cooper, J. C., Horrall, R. M., Hasler, A. D. and Dizon, A. E. (1973). Olfactory hypotheses and salmon migration: a synopsis of recent findings. *Fish Res. Bd Can.* Technical Report. (In press.)

Mainardi, D. and Rossi, A. C. (1970). Chemical communication with regard to the construction of the nest of the anabantid fish, *Colisa colla*. *Ist. Lombardo Acad. Sci. Lett., Rend. Sci. Biol. Med.* **102**, 23–28.

Marshall, N. B. (1971). "Exploration in the Life of Fishes". Harvard University Press.

Mathewson, R. F. and Hodgson, E. S. (1971). Experimental study of amino acids and amines as cues for feeding and aggregations in sharks. *In* "Report of the Director, Lerner Marine Laboratory, American Museum of Natural History, Bimini, Bahamas to Office of Naval Research. Oceanic Biology Branch". Contract No. ONR 552(7), pp. 81–90.

Morrill, A. D. (1899). The pectoral appendages of *Prionotus* and their innervation. *J. Morph.* **11**, 177–192.

Myrberg, A. A. (1966). Parental recognition of young in cichlid fishes. *Anim. Behav.* **14**, 565–571.

Nanba, R., Djahanparwar, B. and von Baumgarten, R. (1966). Erregungsmuster einzelner Fasern des Tractus olfactorius lateralis des Fisches bei Reizung mit verschiedenen Geruchsstoffen. *Pflügers Arch. ges. Physiol.* **288**, 134–150.

Nelson, K. (1964). Behavior and morphology in the glandulocaudine fishes. *Univ. Calif. Publs. Zool.* **75** (**2**), 59–152.

Noble, G. K. and Curtis, B. (1939). The social behavior of the jewel fish, *Hemichromis bimaculatus* Gill. *Am. Mus. Nat. Hist.* **76**, 1–46.

Oshima, K., Gorbman, A. and Shimada, H. (1969). Memory blocking agents: effects on olfactory discrimination in homing salmon. *Science, N.Y.* **165**, 86-88.

Parker, G. H. (1914). The directive influence of the sense of smell in dogfish. *Bull U.S. Bur. Fish.* **33**, 61–68.

Parker, G. H. (1922). "Smell, Taste and Allied Senses in Vertebrates". D. B. Lippincott, Philadelphia.

Parker, G. H. and Sheldon, R. E. (1913). The sense of smell in fishes. *Bull. U.S. Bur. Fish.* **32**, 33–46.

Pfeiffer, W. (1960). Uber die Schreckreaktion bei Fischen und die Herkunft des Schreckstoffes. *Z. vergl. Physiol.* **43**, 578–614.

Pfeiffer, W. (i963). Alarm substances. *Experientia* **19**, 1–11.

Pipping, M. (1926). Der Geruchssinn der Fische mit besonderer Berück sichtigung seiner Bedeutung für das Aufsuchen des Futters. *Soc. Sci. Fennica Commentat. Biol.* **2** (**4**), 1–28.

Pipping, M. (1927). Ergänzende Beobachtungen über den Geruchssinn der Fische mit besonderer Berücksichtigung seiner Bedeutung für das Aufsuchen des Futters. *Soc. Sci. Fennica Commentat. Biol.* **2** (**9**), 1–10.

Poulson, T. L. (1963). Cave adaptation in amblyopsid fishes. *Am. Midl. Nat.* **70**, 257–290.

Price, S. (1969). Chemoreceptor proteins from taste buds. *J. agric. Fd Chem.* **17** (**4**), 709–711.

Rosenblatt, R. H. and Losey, G. S. (1967). Alarm reaction of the top smelt, *Atherinops affinis:* reexamination. *Science, N.Y.* **158**, 671–672.

Roule, L. (1926). "Les Poissons et le Monde Vivant des Eaux". Delagrave, Paris.

Scharrer, E., Smith, S. W. and Palay, S. L. (1947). Chemical sense and taste in the fishes *Prionotus* and *Trichogaster*. *J. comp. Neurol.* **86**, 183–198.

Schemmel, C. (1967). Vergleichende Untersuchungen an den Hautsinnesorganen ober- and unter-irdisch lebender *Astyanax*—Formen. *Z. Morph. Okol. Tiere* **61**, 255–316.

Schnitzlein, H. N. (1968). Introductory remarks on the telencephalon of fish. *In* "The Central Nervous System and Fish Behavior" (D. Ingle, ed.), pp. 97–100. University of Chicago Press, Chicago.

Schutz, F. (1956). Über die Schreckreaktion bei Fischen und die Herkunft des Schreckstoffes. *Z. vergl. Physiol.* **38**, 84–135.

Seitz, A. (1940). Die Paarbildung bei einigen Cichliden. *Z. Tierpsychol.* **4**, 40–84.

Shaw, E. (1970). Schooling in fishes: critique and review. *In* "Development and Evolution of Behavior" (L. R. Aronson, E. Tobach, D. S. Lehrman and J. S. Rosenblatt, eds.), pp. 452–480. Freeman, San Francisco.

Shibuya, T. (1960). The electrical responses of the olfactory epithelium of some fishes. *Jap. J. Physiol.* **10**, 317–326.

Skramlik, E. von (1948). Uber die zur minimalen Erregung des menschlichen Geruchs bzw. Geschmackssinnes notwendigen Molekülmengen. *Pflügers Arch. ges. Physiol.* **249**, 702–716.

Skinner, W. A., Mathews, R. D. and Parkhurst, R. M. (1962). Alarm reaction of the top smelt *Atherinops affinis* (Ayres). *Science, N.Y.* **138**, 681–682.

Steinbach, G. (1950). Funktionelle und vergleichend-anatomische Untersuchungen über Bauchflossen von *Trichogaster trichopterus*, Pallas. *Dt. zool. Z.* **1** (**1**), 24–60.

Steven, D. M. (1959). Studies in the shoaling behavior of fish. *J. exp. Biol.* **36**, 261–280.

Sutterlin, A. M. and Sutterlin, N. (1970). Taste responses in Atlantic salmon *Salmo salar*) parr. *J. Fish. Res. Bd Can.* **27** (**11**), 1927–1942.

Sutterlin, A. M. and Sutterlin, N. (1971). Electrical responses of the olfactory epithelium of Atlantic salmon (*Salmo salar*). *J. Fish. Res. Bd Can.* **28** (**4**), 565–572.

Suzuki, N. and Tucker, D. (1971). Amino acids as olfactory stimuli in freshwater catfish, *Ictalurus catus* (Linn.). *Comp. Biochem. Physiol.* 40A(2): 399–404.

Tavolga, W. N. (1956). Visual, chemical and sound stimuli in sex-discriminatory behavior of the gobiid fish *Bathygobius soporator*. *Zoologica, Sci. Contr. N.Y. Zool. Soc.* **41**, 49–64.

Teichmann, H. (1954). Vergleichende Untersuchungen an der Nase der Fische. *Z. Morph. Ökol. Tiere* **43**, 171–212.

Teichmann, H. (1959). Über die Leistung des Geruchssines beim Aal (*Anguilla anguilla* L.). *Z. vergl. Physiol.* **42**, 206–254.

Teichmann, H. (1962). Was leistet der geruchssinn bei fischen? Umschau, **62** (19), 588–591.

Tester, A. (1963). The role of olfaction in shark predation. *Pacif. Sci.* **17**, 145–170.

Thines, G. (1955). Les poissons aveugles. *Annls. Soc. r. zool. Belg.* **86**, 1–128.

Thines, G. and Vandenbussche, E. (1966). The effects of alarm substance on the schooling behaviour of *Rasbora heteromorpha* Duncker, in day and night conditions. *Anim. Behav.* **14**, 296–302.

Thomas, S. E. (1970). Natural Products from Marine Invertebrates. Ph.D. Thesis. University of the West Indies, Kingston, Jamaica, W.1.

Thorpe, W. H. (1956). "Learning and Instinct in Animals". Methuen, London and Oxford.

Todd, J. H. (1968). The social behavior of the yellow bullhead, *Ictalurus natalis*. Ph.D. Thesis, University of Michigan, Ann Arbor.

Todd, J. H. (1971). The chemical languages of fishes. *Scient. Am.* **224** (**5**), 98–108.

Todd, J. H., Atema, J. and Bardach, J. E. (1967). Chemical communication in social behavior of a fish, the yellow bullhead (*Ictalurus natalis*). *Science, N.Y.* **158**, 672–673.

Tucker, D. and Suzuki, N. (1972). Olfactory responses to Schreckstoff of catfish, olfaction and taste IV. *In* "Proceedings of the Fourth International Symposium, Starnberg, Germany, August 2–4, 1971" (D. Schneider, ed.), pp. 121–127. Wissenschaftliche Verlagsgesellschaft MBH, Stuttgart.

Uchihashi, K. (1953). Ecological study of the teleosts in relation to the brain morphology. *Jap. Reg. Fish. Lab.* Report No. **2**.

Ueda, K., Hara, T. J. and Gorbman, A. (1967). Electroencephalographic studies on olfactory discrimination in adult spawning salmon. *Comp. Biochem. Physiol.* **21**, 133–143.

Whitear, M. (1965). Presumed sensory cells in fish epidermis. *Nature, Lond.* **208**, 703–704.

Whitear, M. (1971a). The free nerve endings .in fish epidermis. *J. Zool., Lond.* **163**, 231–236.

Whitear, M. (1971b). Cell specialization and sensory function in fish epidermis. *J. Zool. Lond.* **163**, 237–264.

Wisby, W. J. and Hasler, A. D. (1954). Effect of olfactory occlusion on migrating silver salmon (*Oncorhynchus kisutch*). *J. Fish. Res. Bd Can.* **11**, 472–478.

Wrede, W. L. (1932). Versuche über den Artduft der Elritzen. *Z. vergl. Physiol.* **17**, 510–519.

Wunder, W. (1927). Sinnesphysiologische Untersuchungen über die Nahrungsaufnahme bei Verschiedenen Knochenfischarten. *Z. vergl. Physiol.* **6**, 67–98.

3. Interspecies and Intraspecies Chemoreception by Marine Invertebrates

A. M. MACKIE and P. T. GRANT

Institute of Marine Biochemistry (Natural Environment Research Council), Aberdeen AB1 3RA, Scotland, U.K.

I. Introduction

Many interactions between organisms involve chemical agents rather than physical force or contact, and the study of these interactions has been termed chemical ecology. When one organism detects a chemical substance secreted by another organism, of the same or different species, it is tempting to use the term chemical communication. However, as Burghardt (1970) has pointed out, the word communication implies intent and he proposes that it should only be used under certain circumstances—he defines communication as "the phenomenon of one organism producing a signal that, when responded to by another organism, confers some advantage (or the statistical probability of it) to the signaler or his group". For this reason, the general term chemoreception is preferable.

II. Interspecies Chemoreception

Interspecies chemoreception describes the situation where an individual of one species detects the presence of another individual of a different species by a chemical cue. Interspecies chemoreception is involved in food detection by herbivores, scavengers and predators, predator recognition by prey and host detection by commensals and parasites. In almost all cases the receiving organism obtains an adaptive advantage and using the nomenclature of Whittaker and Feeny (1971) the chemicals involved are known as kairomones. Conversely, certain prey organisms have developed a chemical defence against predation and these chemicals are termed allomones. Allomones are also produced which affect organisms in other ways—for example by reducing growth.

A. Food Attractants

The role of chemoreception in the detection of food is described in detail by Lenhoff and Lindstedt in another chapter and will not be discussed here.

B. Predator Recognition and Escape Reactions

Life for the majority of animals is a constant struggle for survival against the environment, starvation and predation by carnivores. By a judicious choice of habitat, an animal can increase its chances of survival but almost inevitably its habitat will overlap that of at least one predator. Where a herbivore and carnivore coexist in the same area, both species can maintain a relatively constant population. This arises from the fact that in general more than one species of prey and predator occur together; in this way, no one prey species is consumed exclusively and no single carnivore has an exclusive right to all food species. Given this situation, the number of individuals of a prey species consumed will depend upon a number of factors, including the preferences of predators and the efficiency of the senses of prey and predator.

All these considerations apply equally well in the marine environment, where animals have evolved diverse defences against predation, ranging from camouflage to the secretion of repellant substances. Several instances of convergent evolution are to be found where widely different prey animals have developed similar or identical responses to predators.

A wide variety of marine organisms display escape or defence reactions to predators and this very diversity may be a reflection of the effectiveness of these responses. Clearly if the responses are even only occasionally successful in preventing predation, responding organisms will have an adaptive advantage over non-responding animals. In general the responses are most effective when the rate of movement of prey is equal to or greater than that of the predator and the majority of the reactions are triggered chemically, either on contact or at a distance, by substances secreted by the predatory organisms.

Before an organism can display an effective escape reaction, it must possess the necessary musculature and nervous system to perform the movements involved in the response. However, the varied nature of these responses means that almost all motile marine invertebrates possess at least the potential to perform escape responses. Thus some animals swim in a series of undulating movements, others eject water in a form of jet propulsion and many others perform various forms of twisting, leaping and somersaulting.

1. *Responses to Starfish*

Some of the most spectacular responses are displayed by scallops in the presence of predatory starfish. These reactions were first described by Fischer (1869) for *Pecten maximus* and since then other species of scallop have been shown to behave in the same way. Bauer (1912) demonstrated that the response was due to a chemical secreted by the starfish and Dakin (1910) suggested that the osphradium was responsible for the detection of the starfish. During the escape response, the scallop draws water into its mantle, then closes the valves suddenly, thus expelling the water violently.

Not all starfish are equally effective in eliciting a response. Thus the scallops *Pecten jacobaeus* and *Chlamys opercularis* respond violently to *Marthasterias glacialis* and much less strongly to other starfish tested (Lecomte, 1953) (Table 1). Lecomte drew a distinction between a swimming response, where the scallop moved through the water, mantle edge forwards, and a jump, where the hinge was foremost. Similarly, Thomas and Gruffydd (1971) found that a sustained swimming response was obtained in the larger *Pecten maximus* only on contact with *Asterias rubens*, *Astropecten irregularis* and *Marthasterias glacialis*, starfish known to prey upon *P. maximus*. Jumping was also obtained on contact with *A. rubens*, *A. irregularis*, *M. glacialis* and *Luidia ciliaris* and with extracts of these starfish. In contrast, Fange (1963) found during his investigations into the toxic substances of starfish that extracts of *A. rubens*, *M. glacialis* and *Henricia sanguinolenta* caused contraction of the isolated radula muscle of *Buccinum undatum*, but only the *Marthasterias* extract induced an escape reaction in *P. maximus*.

Mauzey *et al.* (1968) discussed the effectiveness of the escape responses of *Chlamys* species and reported that, in the Puget Sound region, field predation has only been occasionally observed, although many starfish would eat scallops in the aquarium if the scallops were prevented from escaping.

Table 1. The escape response induced in marine invertebrates by starfish, observed either in the field or under laboratory conditions.

(+): indicates a consistent and strong escape response induced by a starfish.
(±): indicates an occasional response.
(−): indicates that no detectable response has been observed.

Responding organism	Starfish		
	Order Phanerozonia	Order Spinulosa	Order Forcipulata
Scallop (*Chlamys opercularis*); Lecomte (1953)	(+) *Astropecten auriatacus*	(+) *Asterina gibbosa* (+) *Palmipes membranaceus* (±) *Echinaster sepositus*	(+) *Marthasterias glacialis*
Limpet (*Acmaea scutum*); Margolin (1964a)	(−) *Mediaster aequalis* (−) *Dermasterias imbricata* (−) *Luidia foliolata*	(−) *Solaster stimpsoni* (−) *Crossaster papposus* (−) *Pteraster tesselatus* (−) *Henricia leviuscula*	(+) *Pisaster ochraceus* (+) *Pisaster brevispinus* (+) *Leptasterias aequalis* (+) *Evasterias troschellii* (+) *Orthasterias koehleri* (+) *Pycnopodia helianthoides*

Limpet			
(*Diodora aspera*);	(±) *Hippasteria spinosa*	(±) *Patiria miniata*	(+) *Pisaster ochraceus*
Margolin (1964b)	(−) *Mediaster aequalis*	(−) *Solaster stimpsoni*	(+) *Pisaster brevispinus*
	(−) *Dermasterias imbricata*	(−) *Solaster dawsoni*	(+) *Pisaster giganteus*
	(−) *Luidia foliolata*	(−) *Crossaster papposus*	(+) *Leptasterias aequalis*
		(−) *Pteraster tesselatus*	(+) *Evasterias troschellii*
		(−) *Henricia leviuscula*	(+) *Orthasterias koehleri*
			(+) *Pycnopodia helianthoides*
Sea cucumber			
(*Parastichopus californicus*);	(+) *Luidia foliolata*	(+) *Solaster stimpsoni*	(+) *Pycnopodia helianthoides*
Mauzey *et al.* (1968)	(+) *Dermasterias imbricata*	(+) *Solaster dawsoni*	(−) *Pisaster ochraceus*
	(−) *Hippasteria spinosa*	(+) *Solaster endeca*	(−) *Pisaster brevispinus*
	(−) *Mediaster aequalis*	(−) *Crossaster papposus*	(−) *Evasterias troschellii*
	(−) *Gephyreaster swifti*	(−) *Pteraster tesselatus*	(−) *Orthasterias koehleri*
		(−) *Henricia leviuscula*	(−) *Stylasterias ferreri*
Sea anemone			
(*Stomphia coccinea*);	(+) *Hippasteria spinosa*	(−) *Solaster stimpsoni*	(−) *Evasterias troschellii*
Ward (1965a); Yentch and	(+) *Hippasteria phrygiana*	(−) *Solaster dawsoni*	(−) *Leptasterias aequalis*
Pierce (1955)	(+) *Dermasterias imbricata*	(−) *Crossaster papposus*	(−) *Leptasterias foliolata*
	(−) *Mediaster aequalis*	(−) *Pteraster tesselatus*	(−) *Orthasterias koehleri*
		(−) *Henricia leviuscula*	(−) *Pisaster brevispinus*
			(−) *Pisaster ochraceus*
			(−) *Pycnopodia helianthoides*

Other bivalves also display escape responses to predatory starfish. Thus, the cockle *Cardium echinatum* responds to contact with *A. rubens* and *M. glacialis* by protruding its foot and pushing hard against the substrate, thereby making a powerful leap upwards and away from the starfish (Feder, 1967). Ansell (1967) studied the leaping behaviour of cockles and showed that *C. echinatum* and *Laevicardium crassus* responded to *A. rubens* while *C. edule* did not. There appeared to be a relationship between occurrence of escape response and habitat, those bivalves which displayed the response living on coarse substrata, while *C. edule* occurs on sandy shores where few asteroid predators are encountered (Ansell, 1969).

A wide variety of gastropods also exhibit escape responses to predatory starfish. Several species of the mudsnail *Nassarius* show violent leaping or somersaulting reactions on contact with *M. glacialis* and *A. rubens* but display no response to *Crossaster papposus*, which preys upon other starfish (Feder, 1967). *N. vibex* responds violently to *Luidia alternata* but not to *Astropecten* species nor to *Echinaster sentus* (Gore, 1966).

Crisp (1971) investigated the distribution of receptor cells of *Nassarius reticulatus* and found that they were most abundant on the anterior border of the foot, tentacle tips and ventral external surface of the siphon tip. Food extracts only were tested on the receptors, but some at least of them are probably involved in the detection of starfish.

The large whelk *Buccinum undatum* performs a violent escape response when it comes into contact with *M. glacialis* but contact with *A. rubens* has no effect (Feder, 1967). Extracts of both starfish, however, caused the response. Studies on the live animals would indicate that *M. glacialis* continually secretes the substance inducing escape (Feder and Arvidsson, 1967).

In some cases there appears to be a degree of correlation between escape response and probability of predation. Those species of the limpet *Acmaea* which did not show a response to predatory starfish in the aquarium were found in the tidal zone where starfish predation is rare (Bullock, 1953). This difference in response was so pronounced that Bullock used it to differentiate between species morphologically indistinguishable. He also found that forcipulate starfish induced escape reactions while spinulose starfish did not. Margolin (1964a) found a similar situation with *Acmaea* species (Table 1).

Feder (1963) reported that, in tidal pools along the coast of California, gastropods were not consumed by starfish in proportion to their relative abundance. Only *Tegula funebralis* and *Acmaea scabra* were consumed to any extent and the latter animal did not show an escape response to *Pisaster ochraceus*. Barnacles and mussels were, however, the preferred food organisms. Paine (1969) also concluded that the escape response of *Tegula* was relatively ineffective in the field and that the starfish *P. ochraceus* showed a marked preference for *Mytilus* species. This indicates that palatability of different prey organisms is an important factor in determining the degree of predation and lack of predation need not necessarily mean that an effective escape reaction is involved. Castilla (1972) has shown that *Asterias rubens* fed on *Mytilus* were not attracted to oysters. If, however, the starfish were fed on oysters for a month they then responded positively to the "odour" of oysters. This is an example of ingestive conditioning first demonstrated by Wood (1968) with the gastropod *Urosalpinx cineria.*

Szal (1971) has demonstrated that the receptors responsible for the detection of starfish at a distance are situated in the ctenidial leaflets of *Tegula.* Each leaflet possesses a chemoreceptor organ—a slight swelling of the efferent border near the base.

A somewhat different response is observed in certain snails and limpets. Thus, on contact with certain forcipulate starfish the limpet *Diodora aspera* raises its shell and extends a fold of its mantle over the shell, dislodging the tube feet of the starfish (Margolin, 1964b). Spinulose and phanerozone starfish had little or no effect. (Table 1.)

The nudibranch *Tritonia gilberti* has been reported to display an escape reaction to *Pycnopodia helianthoides*, *Luidia foliolata* and *Solaster dawsoni*, starfish known to eat the nudibranch in the aquarium (Mauzey *et al.*, 1968). Willows and Hoyle (1969) have studied the neuronal network which brings about this swimming reaction and have demonstrated that the nerve cells responsible are situated in the pleural ganglia. An impulse lasting approximately 1 sec in these neurones triggers a prolonged discharge (up to 1 min) in other brain cells which control the dorsal and ventral muscles involved in the swimming escape response.

The sea urchin *Strongylocentrotus droebachiensis* performs a characteristic running response to contact with *Marthasterias glacialis* while *Psamechinus miliaris* defends itself with its globiferous pedicellaria (Jensen, 1966). *Asterias rubens*, *Crossaster papposus* and *Solaster endeca* were without effect. On contact with *M. glacialis*, the sea urchin *Arbicia bixula*, which lacks globiferous pedicellaria, defends itself by turning its spines out from the test (Vasserot, 1964). A variation in response with different starfish was also reported by Chia (1969) who showed that *Luidia ciliaris* elicited a response in *Echinus esculantis*, *Psamechinus miliaris* and *Strongylocentrotus droebachiensis*, while *Henricia sanguinolentis* and *Astropecten irregularis* did not.

Mauzey, *et al.* (1968) made some interesting observations on the *Pycnopodia–Strongylocentrotus* interaction. The sea urchins generally kept to the side of the aquarium opposite a specimen of *Pycnopodia*; however, when the starfish was feeding on a clam, the urchins no longer showed this avoidance behaviour, some of them actually remaining in contact with the starfish. Immediately the starfish stopped feeding, the urchins again moved away. The authors did not offer any explanation for this behaviour, but it could be due to several factors. One possibility is that the superposition of clam odour inhibits the response to starfish "fright substance", analogous to the inhibition of the swimming reaction of *Stomphia* by food juice (Ross and Suttod, 1964). Another possibility is that the "fright substance" is not secreten when the starfish is feeding. The starfish *Asterias rubens* executes a rapid escape reaction on contact with the sunstars *Crossaster papposus* and *Solaster endeca* and predation has been observed by both sunstars in Denmark and England (Feder, 1967). Mauzey *et al.* (1968) found that *Solaster dawsoni* caused escape reactions in other individuals of the same species and also in *S. stimpsoni*, *C. papposus*, *Mediaster*, *Dermasterias*, *Pycnopodia* and *Evasterias*, all of which are preyed upon by *S. dawsoni*. In the Puget Sound region, however, neither *C. papposus* nor *S. endeca* fed upon other starfish nor did they cause escape reactions. Both Feder and Mauzey concluded that contact and not distance chemoreception was involved. In contrast, Castilla and Crisp (1970), using a simple Y-tank system, demonstrated that *Asterias rubens* could sense *Crossaster papposus* at a distance and moved downstream when the water-borne odour of *Crossaster* was present.

A "running" response of ophiuroids in the presence of *Luidia foliolata* has been described by Mauzey *et al.* (1968) although in the area of observation the brittle stars were not eaten by this starfish, in marked contrast to *L. sarsi* whose main food is brittle stars (Fenchel, 1965). Feder and Arvidsson (1967) found that an extract of *M. glacialis* caused an escape reaction in the ophiuroid *Ophiothrix fragilis* and Passano (1957) reported that several species of ophiuroid display escape reactions to *Solaster* and *Dermasterias*.

The holothurian (sea cucumber) *Parastichopus californicus* displays a swimming escape response in the presence of several species of starfish (Table 1) (Mauzey *et al.*, 1968). Although *Pycnopodia* was the only starfish to have been observed to feed on *Parastichopus*, all the starfish inducing the escape response fed on other holothurians. Conversely, starfish which did not cause a swimming response were never observed to feed upon holothurians, indicating that this is a true escape response to a predator.

A rather similar swimming response is shown by the sea anemone *Stomphia coccinea* in the presence of *Dermasterias* and other starfish (Yentch and Pierce, 1955; Ward, 1965). Starfish of the orders Spinulosa and Forcipulata were without effect (Table 1). The aboral surface of *Dermasterias* was more effective than other parts of the starfish, and extracts prepared by homogenizing the aboral surface in sea water were as effective as the whole animal in eliciting the reaction (Robson, 1961a). It would appear that the *Dermasterias* chemical has a general effect at the tentacle surface rather than a specific one on receptors. The swimming response was inhibited when the tentacular crown was flooded with a food extract (Ross and Sutton, 1964) and the tendency for *Stomphia's* nematocysts to discharge on contact with food disappeared during the swimming response, this disappearance coinciding with the release of the pedal disc. This would indicate some nervous connection between the pedal disc and the nematocysts.

The reaction would appear to be a true escape reaction since occasional predation by starfish has been observed in the field and starfish causing the swimming response in *Stomphia* frequently eat other species of sea anemone (Mauzey *et al.*, 1968).

2. *Responses to Predatory Molluscs*

Many marine molluscs are voracious predators, feeding mainly on other species of mollusc, and a number of the prey animals display avoidance reactions. The herbivorous gastropods *Melagraphia aethiops* and *Zedilonia digna* respond to several carnivorous gastropods in addition to the starfish *Coscinasterias calamaria* and *Astrostole scabra* (Clark, 1958). There did not appear to be any correlation between the sensitivity to different carnivores and the tidal zone occupied by the animals.

Laurenson (1970) has, however, reported that the snail *Melagraphia aethiops* displays a strong escape reaction to the predatory molluscs *Haustrum haustorum* and *Lepsiella scobina* which are frequently encountered, and much weaker responses to predatory molluscs less often encountered. The escape reaction of *M. aethiops* appears to be effective under natural conditions since the mean ratio of prey organisms *Lunella smaragda*, *Lepsiella scobina* and *Melagraphia aethiops* was 1 : 10 : 9 while the ratio of their predation by *H. haustorum* was 1 : 0·45 : 0·069. Once again, however, ingestive conditioning and palatability may be involved.

The snail *Littorina planaxis* shows a definite avoidance reaction to the predatory intertidal snail *Acanthina spirata*, induced by a chemical associated with the mucus (Peters, 1964) and the substance causing an escape reaction in *Nassarius luteostoma* originates in the palial cavity or foot of *Natica unifasciata* (Gonor, 1965). *Nassarius vibex* displays an escape reaction to the gastropods *Fasciolaria tulipa* and *F. hunteria* and once again the active substance was present in the mucus trails left by the predators (Gore, 1966). Both distance and contact chemoreception was involved and pieces of syphon and epidermis from the ventral surface of the foot caused the response.

The response of *Tegula funebralis* to contact with the predatory snails *Thais emarginata* and *Acanthina spirata* is quite different to its escape response to starfish (Yarnall, 1964). Instead of moving away, the snails crawls towards and over the predator.

The escape reactions of *Strombus gibberulus*, *S. luhuanus* and *Lambis lambis* induced by the presence of *Conus marmorus* and *Aulica vespertilio* were studied by Gonor, (1966) who demonstrated that the response was due to a chemical secreted into the water by the predatory gastropods. Berg (1971, 1972) has studied the development of the escape response of *Strombus maculatus* to *Conus* species in veligers and juveniles and has found that the entire escape reaction is first observed in juveniles 6 weeks past metamorphosis and larger than 2·0 mm shell length. Previous exposure to "cone factor" was not necessary and Berg suggests that development of neural connections determines the age at which the escape response is first capable of being performed.

In the aquarium the scallops *Chlamys opercularis* and *Pecten maximus* have been observed to respond violently to contact with the whelk *Buccinum undatum* and a response was also obtained upon the addition of water in which *Buccinum* had been maintained, indicating that the active substance is secreted into the water (Mackie, unpublished observations). *Buccinum* have been observed to attack *P. maximus* in the field (Dakin, 1912) which suggests that the response is a true escape reaction to a predator.

The sea anemones *Stomphia coccinea* and *Actinostola new species* display a swimming escape reaction on contact with the predatory nudibranch *Aeolidia papillosa* (Robson, 1961b; Ross and Sutton, 1967). A specific chemical appears to be responsible since other non-predatory nudibranchs (*Aeolidiella glauca*, *Armina loveni* and *Tritonia hombergi*) did not cause the response. A substance produced by the anterior area of the foot, in the region of the slime groove, caused the escape response when applied to the column.

Berry (1971) has described a dramatic escape response in the rock lobster *Panulirus homarus* to *Octopus granularis* in South African waters. Whether the response is triggered chemically or visually is not, however, known at present.

3. *Responses to Other Marine Invertebrates*

A few isolated instances of escape responses to predators other than starfish and molluscs have been reported. The sea anemone *Actinostola new species* displays its escape response on contact with the anemone

Stomphia coccinea (Ross and Sutton, 1967). No substance causing swimming in *Actinostola* has, however, been detected in extracts of *Stomphia*, indicating that this may not simply be due to the secretion of a repellant substance by the latter species. As far as we are aware, no investigations have been carried out into the possible role of the nematocysts in the escape reaction.

The sea urchin *Diadema antillarum* responds to the presence of injured animals of the same species and to body fluids of the urchins *Echinometrum lucanter* and *Lytechinus variegatus* (Snyder and Snyder, 1970). Fish predators of *Diadema* were also found to eat the other species of urchins, and the response to the juices therefore gives an adaptive advantage to *Diadema*.

4. *Chemical Nature of Substances Inducing Escape Reactions*

Considering the widespread occurrence of escape reactions, surprisingly little information is available concerning the nature of the substances responsible. This is largely due to the lack of reliable quantitative bioassays for the active substances, which makes it difficult to purify them. Feder and Lasker (1964) reported the partial purification of the active material from the tube feet of *Pisaster ochraceus*, using as test animal the file limpet, *Acmaea limatula*. The material was insoluble in diethyl ether, acetone and petroleum ether, soluble in dimethyl sulphoxide and colloidal in water, and appeared to be associated with protein in the tube feet. The activity remained with the colloidal material when fractionated on a column of Sephadex G-25.

The active substance from *Marthasterias glacialis*, inducing an escape reaction in *Buccinum undatum*, was partially purified by Feder and Arvidsson (1967). The material did not diffuse through a dialysis membrane and gel filtration through a column of Sephadex G-50 only partly separated the active substance from colloidal material. Paper chromatography indicated that the active substance was ninhydrin and Ehrlich negative. Mackie *et al.* (1968) developed a sensitive bioassay for the substances secreted by *Marthasterias glacialis* and *Asterias rubens* and showed that they were saponin-like materials, probably existing as a protein complex in the starfish. These materials had surface-active properties and the escape reaction in *Buccinum* could also be elicited with solutions of synthetic surface-active agents. It has since been shown that the active material from *M. glacialis* is a sulphated steroid glycoside, acid hydrolysis yielding steroids, inorganic sulphate and the sugars glucose, 6-deoxy glucose and 6-deoxy galactose (Mackie and Turner, 1970). The major steroid moiety, marthasterone, possessed a novel Δ^{24}-23-ketone structure in the side chain and has now been fully characterized as 3β, 6α-dihydroxy-5α-cholesta-9(11), 24-diene-23-one (Fig. 1) (Turner *et al.*, 1971). The minor steroid was found to be the 24, 25-dihydro derivative of marthasterone (Fig. 1). The steroid glycoside induces escape reactions in a variety of organisms known to respond to *M. glacialis* (Mackie, 1970). The other saponin of

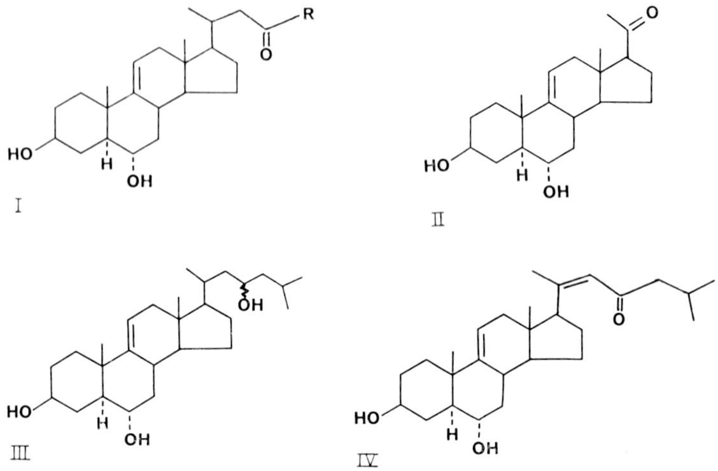

Fig. 1 Structure of steroidal aglycones of starfish saponins.
I, Marthasterone, 3β 6α-dihydroxy-5α-cholesta-9(11), 24-diene-23-one where R is $-CH=C\left\langle\begin{matrix}CH_3\\CH_3\end{matrix}\right.$, and dihydromarthasterone where R is $-CH_2-CH\left\langle\begin{matrix}CH_3\\CH_3\end{matrix}\right.$; II, 3$\beta$, 6$\alpha$-dihydroxy-5$\alpha$-pregn-9(11)-ene-20-one; III, 3β, 6α, 23ξ-trihydroxy-5α-cholest-9(11)-ene; IV, 3β, 6α-dihydroxy-5α-cholesta-9(11), 20(22)-diene-23-one.

M. glacialis was much less active, and synthetic surface active agents also caused a response.

Laurenson (1970) has shown that saponins are the substances responsible for the escape reaction of *Melagraphia aethiops* to *Coscinasterias calamaria*, and Faulkner (personal communication) has confirmed that saponins are responsible for the reactions of molluscs to Californian starfish. We have found that the major steroid glycoside of *Marthasterias* is absent from *Asterias rubens*, *Astropecten irregularis*, *Porania pulvillis* and *Crossaster papposus* (Mackie, unpublished observations). Thus, the unusually violent responses induced by *M. glacialis* would appear to be due to the secretion of a steroid glycoside not found in other starfish.

Saponins have been isolated from several species of starfish (Yasumoto *et al.*, 1966; Rio *et al.*, 1965) and since these starfish are known to cause escape reactions in sensitive animals, it is possible that the saponins are the active substances in all cases. Several of the steroids of these saponins have recently been characterized. Thus, the major steroid of the saponins from *Asterias amurensis* was shown to be 3β, 6α-dihydroxy-5α-pregn-9(11)-ene-20-one (II, Fig. 1) (Ikegami *et al.*, 1972a). Smaller amounts of marthasterone (I, Fig. 1) and 3β, 6α, 23ξ-trihydroxy-5α-cholest-9(11)-ene (III, Fig. 1) were also found (Ikegami *et al.*, 1972b). The saponins of *Acanthaster planci* have been shown to contain 3β, 6α-dihydroxy-5α-cholesta-9(11), 20(22)-diene-23-one (IV, Fig. 1) and compound II (Fig. 1) (Sheikh *et al.*, 1972).

Nothing is known about the biosynthesis of starfish steroid glycosides. However, recent work has shown that starfish steroids are partly dietary in origin and partly synthesized by the animals (Smith and Goad, 1971a, b; Goad *et al.*, 1972). There was a high incorporation of (2-^{14}C)-mevalonic acid into squalene and 4, 4-dimethyl steroids, but the principal steroid component, cholest-7-enol, was only weakly labelled. This would suggest that the latter compound may arise from the modification of dietary steroids and that the 4, 4-dimethyl steroids may give rise to some other steroid constituent—for example, the steroid glycosides (Goad *et al.*, 1972). The $\Delta^{9(11)}$ bond would appear to be characteristic of the steroid glycosides of starfish and sea cucumbers just as the Δ^7 bond is characteristic of the free steroids of these animals (Bergmann, 1962). Another feature common to four of the starfish aglycones is an oxygen at the 23-position, only found previously in 23ξ-hydroxy lanosterol, isolated from the fungus *Scleroderma aurantium* (Entwistle and Pratt, 1968).

Laurenson (1970) has found that urocanylcholine (I, Fig. 2) is the active material causing the escape reaction of the mollusc *Melagraphia aethiops* on contact with the predatory gastropod *Haustrum haustorum*, the substance being produced by the hypobranchial gland. Choline derivatives have been isolated from the hypobranchial glands of several other predatory gastropods, and many of these animals are known to cause escape reactions. Urocanylcholine has been isolated from *Murex* species, *Tritonalia erinacea*, *Urosalpinx cinereus* and *Thais lapillus*, senecioylcholine (II, Fig. 2) from *Thais floridana* and acrylylcholine (III, Fig. 2) from *Buccinum undatum* (Whittaker, 1960).

Ward (1965) investigated the swimming response of *Stomphia* in the presence of *Dermasterias* and concluded that the active substance was an aminopolysaccharide. His active preparation was, however, impure and more work will be necessary before the nature of the substance can be definitely established. Starfish inducing the reaction belong to the order Phanerozonia while those inducing reactions in molluscs are mainly of the order Forcipulata, which would indicate that the substances responsible are different (Table I). In this connection, investigations of the substance from *Crossaster papposus* causing an escape reaction in *Asterias rubens* indicate that it is not a saponin (Mayo and Mackie, unpublished observations). The active material has not yet been fully characterized but at least two components, both of low molecular weight, would appear to be required.

As has already been mentioned, *Crossaster papposus* and *Solaster endeca* feed on other starfish in Denmark and England but not in the

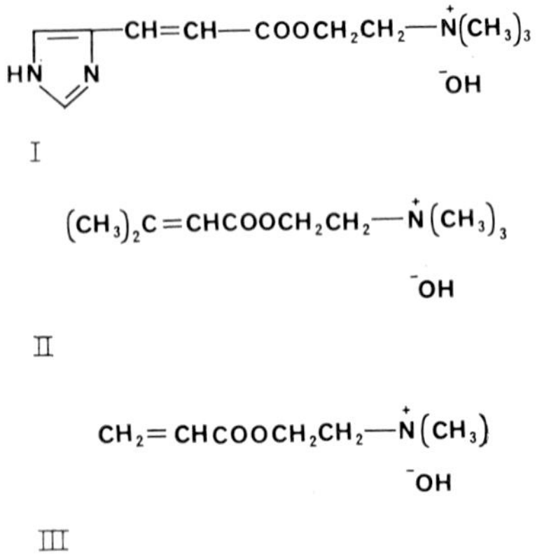

Fig. 2 Structure of choline esters. I, urocanylcholine; II, senecioylcholine; III, acrylylcholine.

Puget Sound area of the United States, and escape reactions were only observed in starfish where predation occurred. In the Puget Sound area, however, *S. dawsoni* eats starfish and causes escape reactions. This would indicate that the repellant substance may be produced by the modification of a compound present in other starfish. Once the nature of the active substance is known it should be possible to confirm or disprove this suggestion.

5. *Evolution of Escape Reactions*

It is generally considered that the escape reactions of gastropod molluscs have evolved from a simple righting reaction (Gore, 1966; Gonor, 1966; Laurenson, 1970) and the leaping movements in bivalves observed on contact with starfish can also be observed when the foot fails to penetrate the substratum (Ansell, 1969). Swimming in forms such as *Pecten* is regarded as a development of a cleaning reflex, normally used to eject irritating particles (Bullock, 1965). Sensitive animals are often found to respond to non-predators as well as predators, indicating a relatively low degree of sensory discrimination. If the escape reactions are due to a small number of chemical compounds, present in many animals, then the predators will in general be indistinguishable from innocuous animals. Differences in the degree and intensity of response of a given prey animal to different predators may be due to two factors —variation in the quantity of repellant substance secreted by the various predators or differences in the nature of the repellants. Conversely, variations in response in different prey organisms to the same predator must be due to differences in the sensitivity of chemosensory cells if all the organisms are in fact responding to the same chemical. Much more sensory discrimination is possible if the prey organism is responding to a mixture of chemicals secreted by a predator, as is the case with *Asterias rubens* and *Crossaster papposus*.

The substances known to induce escape reactions in molluscs—steroid glycosides, urocanylcholine and surface-active agents—all have an effect on muscle and cell membranes. Thus, saponins and synthetic surface-active agents lyse red blood cells, and saponins and urocanylcholine possess neuromuscular blocking activity (Friess, 1972; Erspamer and Glässer, 1957). However, little is known about the neurophysiological and pharmacological effects of these compounds, although the steroid glycosides of *Marthasterias glacialis* are known to affect mammalian and invertebrate smooth muscle (Mackie *et al.*, 1968).

It has been shown that there is a correlation between the sensitivity of some marine invertebrates to synthetic detergents and their reaction to extracts of *Marthasterias* and synthetic surface-active agents (Mackie, 1970). Animals which had the lowest median lethal concentration for the detergent BP 1002 were found to show the most violent reactions to starfish and their extracts. Fischer (1971) reported a similar relationship between an individual animal's taste threshold for a drug and that drug's median lethal dose (LD_{50}), and regarded the latter as "the ultimate systemic pharmacological response". Steroid glycosides and surface-active agents, however, do not come under the definition of a drug which Fischer proposed, but the similarity of the two situations is worthy of further investigations.

The primary function of these starfish steroid glycosides is not known with any certainty. However, their toxicity and repellant properties would suggest that they may serve as defence substances against predators and parasites. At first sight, the secretion of a substance inducing escape reactions in prey animals would appear nonsensical, but if the steroid glycosides serve as a repellant for predators, then it is only to be expected that some prey animals will also be repelled.

There is also some evidence for the involvement of steroid glycosides in the reproductive physiology of starfish. The highest concentration of steroid glycosides occurs in *Asterias amurensis* in the summer, and the lowest in the winter, when this starfish is in breeding condition (Yasumoto *et al.*, 1966). It has been found that the steroid glycosides isolated from the ovaries of *A. amurensis* act as spawning inhibitors in that starfish, thus preventing unnecessary spawning early in the breeding season (Ikegami *et al.*, 1972c).

C. Defensive Secretions

Many molluscs have developed toxic secretions, highly repellant to potential predators. *Lobiger serradifalci* secretes mucus over the whole of its body and this material is repellant to fish and starfish (Stamm, 1968). Some species of the nudibranch *Doridacea* produce an acid secretion which is repellant to fish (Edmunds, 1968). By far the most important source of the secretion in three of the species was the sub-epidermal glands, which are adapted for ejecting fluid away from the animal. These glands possess a muscular wall, narrow bore opening and sphincter muscle. The acid is probably inorganic, but its exact nature is at present unknown.

The opisthobranch *Phyllidia varicosa* secretes a grey toxic mucus upon tactile stimulation, which is toxic to a number of crustacea and to the fish *Mollinesia latipinna*. The toxin was volatile and stable at 95°C. (Johannes, 1963.)

The conch shells *Cassis tuberosa*, *C. madagascarensis* and *C. flammea* can squirt saliva from the proboscis and this saliva is highly toxic to many marine organisms (Cornman, 1963). The animals use this toxic mucus both offensively and defensively and the active principle would appear to be a neurotoxin.

There is some controversy as to whether the sea hare *Aplysia* secretes a repellant substance. Very few marine organisms prey upon *Aplysia*, which indicates that either the sea hare is unpalatable or that it secretes a repellant. The purple fluid ejected when the animal is disturbed is non-toxic and may be analogous to the ink of the octopus (Chapman and Fox, 1969). A thick white fluid is also secreted by sea hares when disturbed, the fluid being produced in the opaline gland (Flury, 1915). This fluid was toxic to a wide range of marine organisms and Kittredge (1971) has reported that the opaline gland does contain a repellant. A toxic substance aplysin, is also found in the digestive gland (Winkler, 1961).

The tropical holothurian (sea cucumber) *Holothuria difficilis* is toxic to fish and can accurately eject its Cuvierian tubules when attacked (Bakus, 1968). The Cuvierian tubules contain the highly toxic substance holothurin, which is effective in repelling or killing predators—Hodgson *et al.* (1967) found evidence for the repulsion of sharks by holothurin under the conditions of their bioassay. The substance holothurin has since been found to be a mixture of saponins or steroid glycosides (Chanley *et al.*, 1959), and one of them has been studied in

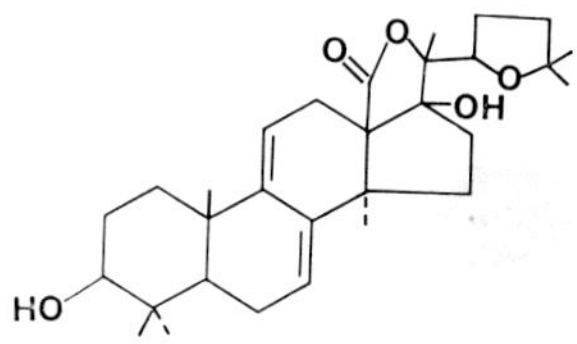

Fig. 3 Structure of 22, 25-oxidoholothurinogenin.

considerable detail. On hydrolysis with aqueous acid, holothurin A from *Actinopyga agassizi* yielded a mixture of reducing sugars, inorganic sulphate and several aglycones (Chanley *et al.*, 1959). The major aglycone was 22, 25-oxidoholothurinogenin (Fig. 3) (Chanley *et al.*, 1966). However, the conjugated diene system found in the aglycones is not present in the glycosides but arises from the acid hydrolysis. It has since been found that the natural aglycones have a double bond in the $\Delta^{9(11)}$-position (Fig. 4) (Chanley and Rossi, 1969).

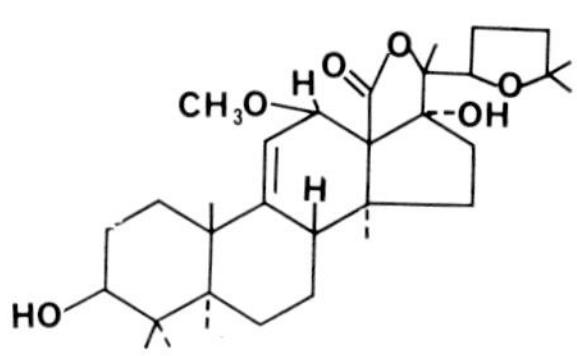

Fig. 4 Structure of one of the neo-holothurinogenins, obtained by enzymic hydrolysis of holothurin A.

In addition to the normal running response, the brittle star *Ophiocomina nigra* also secretes copious quantities of very acidic mucus when attacked (Fontaine, 1964). Histology and chemical analysis indicates that the mucus is rich in highly sulphated mucopolysaccharides, arising in glands in the integument and tube feet.

The secretion of defensive substances is not confined to invertebrates. The mucus of the soapfish *Rypticus saponaceus* contains a haemolytic material which is soluble in water and water-saturated butonal, non-diffusible and moderately stable at 65°C for 2 h (Maretzki and del

Castillo, 1967). The soapfish *Pogonoperca punctata* also secretes mucus containing toxins and thin layer chromatography revealed the presence of several components giving a positive reaction for tertiary amines (Hashimoto and Kamiya, 1969). Further studies suggest that the substances are polypeptides possessing tertiary or quaternary amine moieties (Hashimoto and Oshima, 1972).

Pahutoxin, isolated from the mucus of *Ostracion lentiginosus*, has also been found to be a tertiary amine—the choline ester of 3-acetoxy hexadecanoic acid (Fig. 5) (Boylan and Scheuer, 1967).

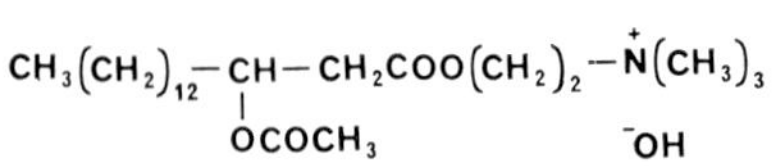

Fig. 5 Structure of pahutoxin.

In conclusion, it can be stated that, although the steroid glycosides of starfish and holothurians and the toxins of soapfish differ in their chemical structure, they all serve a similar purpose and all are haemolytic. Similarly, the choline esters of gastropods and pahutoxin all possess varying pharmacological activities and some of these chemicals may possibly have some medical uses. Baslow (1969, 1971) has reviewed the literature on toxins and other biologically active substances of marine origin and also summarized their pharmacological potential. No doubt, as more work is carried out into the biology and chemistry of marine organisms, many other biologically active compounds will be isolated.

D. Commensalism, Symbiosis and Parasitism

Chemoreception is only one of several factors involved in the establishment of the association between host and symbiont. As Cheng (1967) has pointed out, chemotaxis is generally operative only at short range and before this, the two organisms approach each other either by chance or under the influence of other factors.

Let us first consider the association of the miracidia of *Schistosoma mansoni* with the snail *Australorbis glabratus*—although these are freshwater organisms, much work has been carried out on the association and it is a useful illustration. Etges and Decker (1963) conclusively demonstrated chemotactic attraction of the miracidia in the aquarium; however, they expressed doubt whether such a stimulus is operative under natural conditions and considered that light and gravity are more powerful influences in determining orientation of miracidia.

McInnis (1965) found that short-chain fatty acids, some amino acids and a sialic acid derivative were attractive to miracidia of *S. mansoni* and stimulated attachment to, and penetration of, agar blocks. Extraction of the tissue of *A. glabratus* with distilled water, diethyl ether, acetone and ethyl alcohol removed the attractive substance and subsequent addition of butyric acid to the snail tissue restored the capacity of the tissue to attract.

Cheng *et al.* (1966) reported that the plasma and tissue extracts of seven different species of marine pelecypods had similar effects on the cercaria of *Himasthla quissetensis*, but in varying degrees depending upon the species. This would indicate that the attractant produced by the molluscan hosts is non-species-specific. Cheng (1967) has, however, pointed out that chemotaxis may not always be associated with host specificity, the successful or unsuccessful development of the parasite perhaps being independent of this attraction.

The polychaete *Arctonoë fragilis*, commensal with the starfish *Evasterias troschelli*, is attracted to the host from a distance, being able to distinguish between water passing over the host from sea water alone. Individuals commensal with *Evasterias* are not attracted to the non-host starfish *Pisaster* (Davenport, 1950). The host factor from *Evasterias* is non-diffusible and damage to the host starfish abolishes attraction, thus ensuring association with only healthy hosts (Davenport and Hickok, 1951). Three populations of *A. fragilis* (one commensal with *Evasterias*, another with *Orthasterias* and the third with *Solaster*) have been investigated and each population shown to respond to host starfish only (Hickok and Davenport, 1957). Several populations of *A. pulchra*, commensal with various species of starfish, sea cucumber and limpet, were examined for their ability to detect chemical signals (Dimock and Davenport, 1971). Four of the five populations exhibited what was regarded as host recognition.

Rather less specificity was shown by the polychaete *Acholoë astericola*, commensal with *Astropecten irregularis*, and in this association there did not appear to be any distance chemoreception (Davenport, 1953). *A. astericola* responded to the tube feet of the host starfish and to those of *Luidia* and, to some extent, of *Porania*, *Asterina*, *Palmipes* and *Henricia*. None of the forcipulate starfish tested were effective. All parts of the starfish were attractive, particularly the stomach. Contact with a surface, to which the attractant was firmly bound, was necessary for a normal response to be elicited, and no attractant could be obtained in a soluble form.

The facultative commensal polychaete *Podarke pugettensis* is attracted to the starfish *Luidia* and *Patiria* and the attractant has been obtained in a soluble form (Davenport *et al.*, 1960). The attractant was stable in solution at 20°, non-diffusible and destroyed by heating at 90°–95° for 20 min and on incubation with the proteolytic enzyme, trypsin, suggesting that the attractant was a protein. Stewart (1970) has shown that the chemotactic response of *Ophiodromus pugettensis*, commensal with *Patiria miniata*, is host specific. The attractant produced by *Patiria* consists of a small molecular weight substance bound to a larger carrier molecule and the active material can be released by treatment with 4M urea. Gel chromatography on Sephadex G-15 resolved the attractant into three components, two of which are required for activity. Stewart (1970) concluded that the response of *O. pugettensis*, commensal with *Petalaster* (= *Luidia*) *foliolata*, is not strictly host specific and the attractant is of a different type to that of *Patiria*.

The flatworm *Bdellowa candida* is attracted to its host, *Limulus polyphemus*, by a substance which is stable in sea water for at least 5 days. The substance is diffusible through a dialysis membrane and boiling for 15 min destroys the activity (Chevalier and Steinbach, 1969). The effect was species specific since the flatworm was not attracted by water bathing *Libinia*, *Cancer* or *Pagurus*.

Adult specimens of the platyhelminth skin parasite *Entobdella soleae* are found exclusively on the common sole *Solea solea* and laboratory experiments have confirmed that in the dark, larvae of *E. soleae* attached to the skin of *S. solea* in preference to the skin of other soleid and pleuronectid fish (Kearn, 1967). Agar jelly which had been in contact with *S. solea* skin also attracted the larvae, indicating that an attractant substance was produced by the skin. Further experiments showed that the parasites were attracted to host epidermis containing mucus cells but not to host cornea which contained no mucus cells, suggesting that the attractant originates in the epidermal mucus cells.

The commensal crabs *Pinnotheres maculatus* are attracted to their hosts, the bay scallop *Aequipecten irradians concentricus* and the penshell *Atrina rigida* (Sastry and Menzel, 1962). Crabs removed from bay scallops showed no preference for one host over the other. Experiments with the obligate commensal crab *Pinnixa chaetopterana* indicate that the crab is attracted to its host parchment worm *Chaetopterus pergamentacus* by means of a relatively stable protein (Davenport *et al.*, 1960).

The amphipod crustacean *Listriella clymenellaeis* is commensal in the tubes of the polychaete *Clymenella troquata*, and experiments suggest that substances produced by the polychaete, associated with the mucus, are important in maintaining the association (Batcheller and Mills, 1965). However, the response was relatively non-specific since the amphipod responded positively to mucus of *Pectinaria gouldi* as well as to that of *Clymenella*. The shrimps *Betaeus hartfordi* and *B. macginnitae*, commensal with *Haliotis* species and *Strongylocentrotus* species respectively, appear to show a degree of specificity to stimulation by chemicals secreted by living animals (Ache and Case, 1969). Action potentials were monitored from the antennules of *B. macginnitae* when these were stimulated by effluents from the natural hosts, the response of effluents from non-hosts being much less. Glutamate and glycine were also excitory but not betaine and trimethylamine.

The attraction of the commensal bivalves *Montacuta substriata* and *M. ferruginosa* to their respective hosts, the spatangoids (sea urchins) *Spatangus purpureus* and *Echinocardium cordatum*, is relatively non-specific (Gage, 1966a). *M. substriata* is also found with *E. flavescens* or *E. pennatifidus* and *M. ferruginosa* with *S. purpureus*, *E. flavescens* and *E. pennatifidus*. Both species of *Montacuta* are highly motile when isolated from their hosts and this behaviour was analysed in a trough apparatus. The attraction to the host was again found to be non-specific, both species responding to any of the spatangoids, regardless of the species of urchin from which the bivalve had been removed (Gage, 1966b).

A large number of commensal associations between hermit crabs and hydroids and sea anemones have been reported in the literature. The hermit crab *Pagurus bernhardus* is often associated with the anemone *Calliactis parasitica*. Experiments have shown that the anemone responds to shells of living molluscs, and to empty shells, just as readily as to shells in which crabs had lived (Ross, 1967). The "shell factor" which induces the reaction in *Calliactis* has not yet been identified but it is known to be an organic compound present in the periostracum, the horny outer layer of the shell. This substance is a stable compound, insoluble in acetone, ether and alcohol, but dissolved by alkali (Ross and Sutton, 1961a).

Unlike *P. bernhardus*, another crab *Dardanus arrosor* is often active in establishing its association with the anemone *Calliactis* (Ross and Sutton, 1961b). Not all individuals of *Dardanus* were, however, active and the population could be divided into "performers" and "non-performers". The latter behaved like *P. bernhardus* and in this case *Calliactis* transferred to the shell unaided. The performers, who comprised about half the population, participated actively in transferring the anemone to their shells.

Davenport *et al.* (1961) noted that when the base of *Calliactis* was on a shell, its tentacles did not stick to another shell. The variation in the responses of tentacles to the shell factor, when the base was attached to a shell or some other surface, indicated some type of information transfer from base to tentacles. Since Davenport *et al.* (1961) also showed that the nematocysts are involved in the clinging response, it would appear that this "feedback" mechanism may have something in common with the disappearance of nematocyst discharge on contact with food, observed when the pedal disc of *Stomphia* is released prior to the swimming response (Ross and Sutton, 1964).

The association between hermit crabs and sea anemones has an obvious advantage for the latter since they have a greater opportunity to obtain food particles. In only one case, however, has it been definitely shown that the association is of advantage to the crab. This is the *Dardanus-Calliactis* association where it has been found that the presence of the anemone completely stops predation on the crab by the octopus *Octopus vulgaris* (Ross, 1971). Perhaps it is more than mere coincidence that in this association the hermit crab actively encourages the attachment of the sea anemone. Wright (1973) has also shown that hydroid colonies protect their *Pagurus* hosts from another crab, *Clibanarius vittatus* and probably from predators.

III. Intraspecies Chemoreception

Chemical communication has long been known to be involved in many aspects of the life of insects—sex attraction, trail marking and species recognition (Butler, 1967; Wilson, 1968). The term pheromone was first proposed by Karlson and Luscher (1959) to describe a substance which is "secreted to the outside of an individual and received by a second individual of the same species, in which it releases a specific reaction, for example, a definite behaviour or a developmental process". It has recently been shown that the vaginal secretions of Rhesus monkeys contain short chain fatty acids which stimulate sexual behaviour in males of the same species (Michael *et al.*, 1971). With pheromones involved in two such widely different phyla, one would expect them also to be involved in at least some marine organisms.

In animals which discharge their gametes into the sea it is essential for fertilization of the eggs that eggs and sperm are released at the same time. Hence, the production of a "spawning inducer" will obviously be to the advantage of the organisms. Similarly, those animals whose reproduction involves copulation require the presence of ripe males and females in the same area at the appropriate time and sex attractants are produced by many species to accomplish this.

Airborne pheromones all possess certain common characteristics—volatility, molecular weights in the range 80 to 300 and a carbon number of between 5 and 20 (Wilson and Bossert, 1963). In an aqueous environment, volatility is not important and the pheromones of marine organisms would probably have an entirely different nature.

A. Gamete Attractants

Attractants for gametes have been detected in a number of lower marine organisms. Dan (1950) found evidence for a specific attractant for sperm in the eggs of the medusan *Spiracodon saltatrix*, and the female gonangia of the hydroid *Campanularia flexuosa* produce an attractant for sperm which is species specific (Miller and Nelson, 1962; Miller, 1963, 1964, 1965). The chemical nature of these attractants has not been investigated.

In the water mold *Allomyces* the female gametes produce the sperm-attracting substance sirenin which has been characterized as a sesquiterpene (Fig. 6) (Machlis *et al.*, 1968). The sperm-attracting sub-

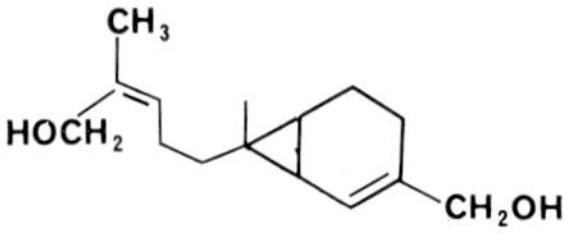

Fig. 6 Sirenin, gamete attractant of *Allomyces*.

stance produced by the female gametes of the marine alga *Ectocarpus siliculosus* has been identified as allo-*cis*—(cycloheptadiene-2′, 5′-yl)-butene-1 (Fig. 7) (Müller *et al.*, 1971). Gamete attractants are also

Fig. 7 Allo-cis-(cycloheptadiene-2′, 5′-yl)-butene-1; gamete attractant of *Ectocarpus*.

produced by *Fucus* species and the gamete attractant of *F. serratus* has been characterized as 1,3,5-octatriene (fucoserraten) (Müller and Jaenicke, 1973).

There is no evidence for chemotaxis of spermatozoa towards eggs in higher organisms, and the only instances in terrestrial organisms are to be found in lower plants such as ferns, mosses and liverworts (Rothschild, 1956).

B. Spawning Inducers

Epidemic spawning occurs in many species of marine invertebrate and a number of factors are involved. Galtsoff (1940) showed that in the oyster *Ostrea virginica*, spawning can be induced in both sexes by raising the temperature of the water, by chemical stimulation or by a combination of both. In the case of the male, the effect was non-specific since spawning could be induced by foreign eggs and a wide range of chemicals, in addition to eggs of *O. virginica*. The females, however, only responded to sperm from males of the same species and the active material was extractable with alcohol and benzene (Galtsoff, 1938). Galtsoff also showed that simultaneous spawning in a large oyster population was induced by the spawning of a few individuals and this effect has been utilized commercially to induce spawning in *O. gigas* (Quayle, 1940).

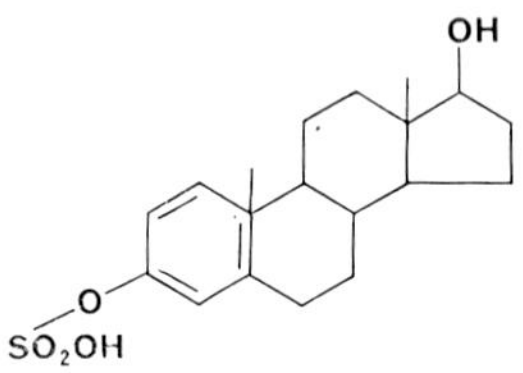

Fig. 8 Oestradiol 3-sulphate.

In the starfish *Asterias rubens*, spawning is induced in ripe specimens of either sex by the presence in the water of the gametes of the other sex (Gemmill, 1914). The presence of a spawning *Paracentrotus lividus* induces other sea urchins of the same species to spawn and water containing eggs or sperm also causes ripe individuals to spawn (Fox, 1924). Creange and Szego (1967), during investigations into the metabolism of oestradiol, showed that the urchin *Strongylocentrotus franciscanus* synthesized oestradiol 3-sulphate (Fig. 8) which was excreted via the gut. They postulated that this highly water-soluble conjugate may have pheromonal activity.

Chaet (1966a) demonstrated that extracts of starfish radial nerves caused isolated gonads to shed their gametes; some species specificity was found. The substance responsible (gonad stimulating substance, GSS) was later shown to be a polypeptide, acting on the ovarian wall to induce the release of another compound (meiosis inducing substance, MIS) (Kanatani, 1967). This latter substance has been characterized as 1-methyl adenine (Fig. 9) (Kanatani *et al.*, 1969).

Chaet (1966b) postulated that GSS might act as a pheromone, stimulating gamete release in neighbouring starfish. However, Kanatani and Shirai (1968) have shown conclusively that neither MIS nor GSS acts as a pheromone, both substances having to be injected into the coelomic cavity of the starfish for any effect to be produced. As has already been mentioned, the presence of eggs or sperm in the water induces spawning in ripe individuals. It would therefore seem likely that any pheromone must be released at the same time as the gametes, but as far as we are aware no work has been carried out on this problem. The situation may be analogous to that in ascidians, where there is a hormonal and neural control of gamete release (Carlisle, 1951). These organisms respond to

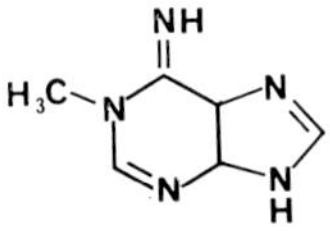

Fig. 9 1-Methyl-adenine.

eggs or sperm of their own species drawn in with the water current and it is suggested that the neural gland is stimulated to secrete gonadotrophin.

C. Sex Attractants and Other Pheromones

Chemicals are involved in the sexual attraction and differentiation of molluscs. In snails of the genus *Crepidula*, association of sexually undifferentiated young with females is common at an early age, and this association prolongs the initial male stage (Coe, 1953). A substance is liberated which attracts the young to her vicinity, and the effect is species specific. A pheromone is also produced by females of the limpet *Littorina planaxis* which attracts males of the same species (Peters, 1964).

Sexual dimorphism in the annelid *Bonellia fulginosa* is under chemical control. Larvae develop into females if they settle on an inert support, but if they settle on a female of the same species they develop into smaller parasitic males (Nigrelli *et al.*, 1967). The substance thought to be responsible, bonellinin, is toxic to a wide range of organisms and cells (Nigrelli, 1963; Nigrelli *et al.*, 1967).

Perhaps the firmest evidence for sex attraction in marine organisms has been obtained with crustacea. Forster (1951) reported that in *Leander serratus*, contact chemoreception is involved in mating. If an "attractive" female was placed in a tank with males, nothing happened until the antennae of a male touched part of the female. The male then swam in circles rapidly until he again made contact. A similar situation occurs with *Pandalus borealis* (Carlisle, 1959).

Brockhuysen (1936) reviewed the literature on the mating behaviour of the shore crab *Carcinides maenas* and concluded that only the sense of touch was important in finding a mate. In contrast, Edwards (1966) reported a strong attraction between the sexes in the edible crab *Cancer pagurus* prior to the female casting her shell, and Ryan (1966) has demonstrated the release of a sex attractant by premolt females of the crab *Portunus sanguinolentis*. Plugging the secretory pores of the

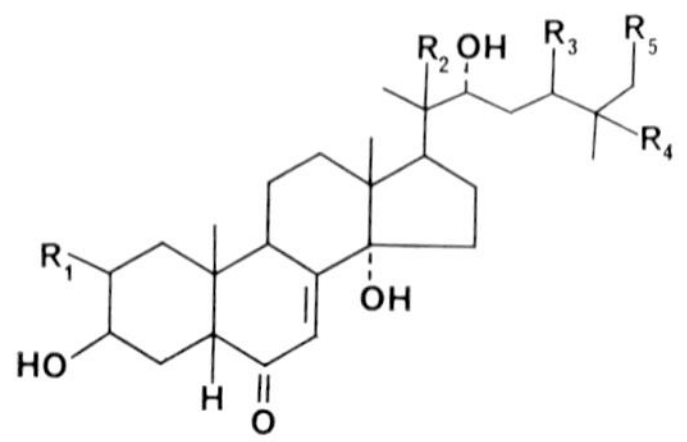

	R_1	R_2	R_3	R_4	R_5
Crustecdysone	OH	OH	H	OH	H
2-Deoxycrustecdysone	H	OH	H	OH	H
Callinecdysone A	OH	OH	H	H	OH
Callinecdysone B	OH	OH	CH_3	OH	H

Fig. 10 Structures of crustacean molting hormones.

female abolished attraction. The effect was species specific—male *P. sanguinolentis* did not respond to *Thalamita crenata* or *Podophelmas vigil* premolt females.

Christofferson (1970) carried out electrophysiological and chemical investigations of the female sex attractant of *P. sanguinolentis* and found that the substance possessed a nett positive charge between pH 2 and 10·85. The pK appeared to lie between 4·0 and 7·25 and the molecular weight was estimated to be 1000 or less.

Kittredge *et al.* (1971) reported that males were attracted to premolt females in the crabs *Pachygrapsus crassipes*, *Cancer antennarius* and *C. anthonyi* and there appeared to be little or no species specificity involved. The molting hormone crustecdysone (Fig. 10) mimiced the effect of the sex pheromone in inducing the premating stance in males of all three species. The threshold concentrations for this effect were 10^{-13}M for *P. crassipes*, 10^{-10}M for *C. antennarius* and 10^{-8}M for *C. anthonyi*. The sex attractant secreted by the female crabs had identical chromatographic properties to crustecdysone, indicating that the two compounds are identical or similar in molecular structure. Kittredge *et al.* (1971) suggest that the species specificity observed by Ryan (1966) with *P. sanguinolentis* may be due to the fact that some species may respond to other molting hormones such as 2-deoxycrustecdysone or callinecdysones A and B (Fig. 10), although none of these compounds were tested. This would, however, appear unlikely from the properties of the *P. sanguinolentis* sex attractant described by Christofferson (1970), unless the molting hormone occurred in a conjugated form.

McLeese (1970) found that male lobsters (*Homarus americanus*) responded more frequently to water from a tank containing a recently molted female than to water from a tank containing a hard female or molted or hard male, indicating that the molting female may release a sex attractant. Atema and Engstrom (1971) studied the responses of male lobsters in greater detail and confirmed that the male responded to the "odour" of a molted female in a different manner than to that of a molted male. Females which had been induced artificially to molt by eyestalk removal did not elicit this sexual response in the male.

Dahl *et al.* (1970) showed that when tritium-labelled females of the amphipod *Gammarus duebeni* were put into one compartment of an aquarium partitioned by a wall of double nylon net, and males placed in the other compartment, the males became excited and swam to the nylon net. Radioactivity was detected in the second antennae of the males, mainly in the calceoli, structures found only in the males. This would indicate that a sex attractant is released into the water by the females. Control experiments with labelled males were, unfortunately, not carried out to eliminate the possibility of non-specific labelling.

When the term pheromone was first introduced, a relationship was implied with hormones, pheromones being considered as external hormones. Conversely, Haldane (1955) reasoned that, since protozoa evolved before metazoa, interindividual chemical signals gave rise to internal hormones. This relationship has recently been considered in detail in Arthropoda by Kittredge and Takahashi (1972), who concluded that sex attractants arose from the molting hormones of primordial crustacea by externalization of the chemoreceptor sites and the controlled release of the hormone into the external environment.

Schooling in *Nassarius obsoletus* is also under chemical control (Crisp, 1969). Water-borne substances are involved in addition to mucus trails laid by other individuals. These trails are polarized in some way since an animal is able to determine the direction in which the trail was laid. The snail *Littorina planaxis* also follows mucus trails laid by others of the same species (Peters, 1964). The cephalic tentacles appear to be responsible for detection and trails laid by other species of mollusc were much less effective. Homing in the pulmonate limpet *Siphonaria alternata* is effected by retracing mucus trails previously laid—no use is made of topographical memory or reverse-displacement (Cook, 1971).

Growth inhibiting substances are fairly widespread in plants (for a recent review see Went, 1970), and there is some evidence for their occurrence in lower animals. The fresh-water snail *Biomphalaria sudanica* produces a compound with growth-inhibiting properties (Berrie and Visser, 1963). This substance has been identified as a monohydroxy tricarboxylic acid monoisodecyl dimethyl ester and was lethal when present in water at twice the normal concentration.

IV. Concluding Remarks

The importance of chemoreception for marine invertebrates cannot be overemphasized, and with the increasing awareness of the dangers of pollution to marine life, it has become vital for as much information as possible on this subject to be obtained. Chemosensory cells are by their very nature extremely sensitive to the presence of low concentrations of chemicals in the environment, and should these chemicals be toxic at higher concentrations, their presence at low concentrations may have a dramatic effect on the behaviour or physiology of organisms.

Pollutants can either cause a behavioural response themselves or they may interfere with normal chemoreception. Thus, small amounts of surface active agents cause escape reactions in molluscs and brittle stars, identical to those observed with starfish or their steriod glycosides. However, prolonged maintenance of the animals in low concentrations of surface active agents results in the almost complete abolition of escape reactions to subsequent stimulation with surface active agents (Mackie, 1970). In this condition, such animals would fall easy prey to predatory starfish.

Low concentrations of heavy metals can attract crustacea and this could bring the animals into lethal concentrations (Portmann, 1972; Mackie and Shelton, 1972). It has also been shown that lobsters are attracted by hydrocarbons (Blumer, 1970). In contrast, water soluble fractions of kerosene interfere with food detection by the snail *Nassarius obsoletus* (Jacobson and Boylan, 1973).

Chemoreception is involved, along with other factors, in site selection and settlement of invertebrate larvae and is discussed by Crisp in Chapter 5. Pollution could well interfere with this behaviour, thus preventing the settlement of commercially important species such as scallops and oysters.

References

Ache, B. and Case, J. (1969). Analysis of antennular chemoreception in two commensal shrimps of the genus *Betaeus*. *Physiol. Zool.* **42**, 361–371.

Ansell, A. D. (1967). Leaping and other movements in some cardiid bivalves induced by *Asterias rubens*. *Anim. Behav.* **15**, 421–426.

Ansell, A. D. (1969). Leaping movements in the Bivalvia. *Proc. malac. Soc. Lond.* **38**, 387–399.

Atema, J. and Engstrom, D. G. (1971). Sex pheromone in the lobster, *Homarus americanus*. *Nature, Lond.* **232**, 261–263.

Bakus, G. J. (1968). Defensive mechanisms and ecology of some tropical holothurians. *Mar. Biol.* **2**, 23–32.

Baslow, M. H. (1969). "Marine Pharmacology: A Study of Toxins and Other Biologically Active Substances of Marine Origin". Williams and Wilkins, Baltimore.

Baslow, M. H. (1971). Marine toxins. *A. Rev. Pharmac.* **11**, 447–454.

Batcheller, R. and Mills, E. L. (1965). Behavioral studies on the commensal amphipod crustacean *Listriella clymenella* Mills. *Biol. Bull. mar. biol. Lab. Woods Hole* **129**, 398.

Bauer, V. (1912). Zur Kenntnis der Lebenweise von *Pecten jacobaeus* L. *Zool. Jb. Abt. 3, Allg. Zool. Physiol. der Tiere* **33**, 127–150.

Berg, C. J. Jr. (1971). A comparative and ontogenetic study of the behavior of strombid gastropods. Ph.D. Thesis, University of Hawaii.

Berg. C. J. Jr. (1972). Ontogeny of the behavior of *Strombus maculatus* (Gastropoda: Strombidae). *Am. Zool.* **12**, 427–443.

Bergmann, W. (1962). Sterols: their structure and distribution. *In* "Comparative Biochemistry" (M. Florkin and H. S. Mason, eds.), Vol. 3, pp. 103–162. Academic Press, New York and London.

Berrie, A. D. and Visser, S. A. (1963). Investigations of a growth-inhibiting substance affecting a natural population of freshwater snails. *Physiol. Zool.* **36**, 167–173.

Berry, P. F. (1971). The spiny lobsters (Palinuridae) of the east coast of Southern Africa: distribution and ecological notes. *Invest. Rep. Oceanogr. Res. Inst.* **27**, 1–23.

Blumer, M. (1970). Oil contamination and the living resources of the sea. *In* "Technical Conference on Marine Pollution and its Effects on Living Resources and Fishing". F.A.O., Rome.

Boylan, D. B. and Scheuer, P. J. (1967). Pahutoxin: a fish poison. *Science, N.Y.* **155**, 52–56.

Brockhuysen, G. J. Jr. (1936). "On the Development and Distribution of *Carcinides maenas* (L.)". J. Enschede, Harlem.

Burghardt, G. M. (1970). Defining "communication". *In* "Advances in Chemoreception, Vol. I—Communication by Chemical Signals" (J. W. Johnston, D. G. Moulton and A. Turk, eds.), pp. 5–18. Appleton-Century-Crofts, Meredith Corp., New York.

Butler, C. G. (1967). Insect pheromones: a review. *Biol. Rev.* **42**, 42–87.

Bullock, T. H. (1953). Predator recognition and escape responses of some intertidal gastropods in presence of starfish. *Behaviour* **5**, 130–140.

Bullock, T. H. (1965). Mollusca: Pelecypoda and Scaphopoda. *In* "Structure and Function in the Nervous Systems of Invertebrates" (T. H. Bullock and G. A. Horridge, eds.), Vol. II, pp. 1388–1431. W. H. Freeman and Co., San Francisco and London.

Carlisle, D. B. (1951). On the hormonal and neural control of the release of gametes in ascidians. *J. exp. Biol.* **28**, 463–472.

Carlisle, D. B. (1959). On the sexual biology of *Pandalus borealis* (Crustacea, Decapoda). II. The Termination of the male phase. *J. mar. biol. Ass. U.K.* **38**, 481–491.

Castilla, J. C. (1972). Responses of *Asterias rubens* to bivalve prey in a Y-maze. *Mar. Biol.* **12**, 222–228.

Castilla, J C. and Crisp, D. J. (1970). Responses of *Asterias rubens* to olfactory stimuli. *J. mar. biol. Ass. U.K.* **50**, 829–847.

Chaet, A. B. (1966a). Neurochemical control of gamete release in starfish. *Biol. Bull. mar. biol. Lab. Woods Hole* **130**, 43–58.

Chaet, A. B. (1966b). The gamete-shedding substances of starfishes; a physiological-biochemical study. *Am. Zool.* **6**, 263–271.

Chanley, J. D. and Rossi, C. (1969). The neo-holothurinogenins- III Neo-holothurinogenins by enzymatic hydrolysis of desulphated holothurin A. *Tetrahedron* **25**, 1911–1920.

Chanley, J. D., Ledeen, R., Wax, J., Nigrelli, R. F. and Sobotka, H. (1959). Holothurin. I. The isolation, properties and sugar components of holothurin A. *J. Am. chem. Soc.* **81**, 5180–5183.

Chanley, J. D., Mazzetti, T. and Sobotka, H. (1966). The holothurinogenins. *Tetrahedron* **22**, 1857–1884.

Chapman, D. J. and Fox, D. L. (1969). Bile pigment metabolism in the sea-hare *Aplysia*. *J. exp. mar. Biol. Ecol.* **4**, 71–78.

Cheng, T. C. (1967). Marine molluscs as hosts for symbiosis: with a review of known parasites of commercially important species. *Adv. mar. Biol.* **5**, 1–424.

Cheng, T. C., Shuster, C. N. and Anderson, A. H. (1966). Effects of plasma and tissue extracts of marine pelecypods on the cercaria of *Himasthla quissetensis*. *Expl. Parasit.* **19**, 9–14.

Chevalier, R. L. and Steinbach, H. B. (1969). A chemical signal attracting the flatworm *Bdellowa candida* to its host, *Limulus polyphemus*. *Biol. Bull. mar. biol. Lab. Woods Hole* **137**, 394.

Chia, Fu-Shiang (1969). Responses of the globiferous pedicellariae to inorganic salts in three regular echinoids. *Ophelia* **6**, 203–210.

Christofferson, J. P. (1970). An electrophysiological and chemical investigation of the female sex pheromone of the crab *Portunus sanguinolentus* (Herbst). Ph.D. Thesis, University of Hawaii.

Clark, W. C. (1958). Escape responses of herbivorous gastropods when stimulated by carnivorous gastropods. *Nature, Lond.* **181**, 137–138.

Coe, W. R. (1953). Influences of association, isolation and nutrition on the sexuality of snails of the genus *Crepidula*. *J. exp. Zool.* **122**, 5–19.

Cook, S. B. (1971). A study of homing behaviour in the limpet *Siphonaria alternata*. *Biol. Bull. mar. biol. Lab. Woods Hole* **141**, 449–457.

Cornman, I. (1963). Toxic properties of the saliva of *Cassis*. *Nature, Lond.* **200**, 88–89.

Creange, J. E. and Szego, C. M. (1967). Sulphation as a metabolic pathway for oestradiol in the sea urchin *Strongylocentrotus franciscanus*. *Biochem. J.* **102**, 898–904.

Crisp, M. (1969). Studies on the behavior of *Nassarius obsoletus* (Say) (Mollusca, Gastropoda). *Biol. Bull. mar. biol. Lab. Woods Hole* **136**, 355–373.

Crisp, M. (1971). Structure and abundance of receptors of the unspecialized external epithelium of *Nassarius reticulatus* (Gastropoda, Prosobranchia). *J. mar. biol. Ass. U.K.* **51**, 865–890.

Dahl, E., Emanuelsson, H. and Macklenburg, G. von (1970). Pheromone transport and reception in an amphipod. *Science, N.Y.* **170**, 739–740.

Dakin, W. J. (1910). The visceral ganglion of *Pecten*, with some notes on the physiology of the nervous system. *Mitt. zool. Stn. Neapel* **20**, 1–40.

Dakin, W. J. (1912). Buccinum (The Whelk). *L.M.B.C. Mem. typ. Br. mar. Pl. Anim.* **20**.

Dan, J. C. (1950). Fertilization in the medusan *Spiracondon saltatrix*. *Biol. Bull. mar. biol. Lab. Woods Hole* **99**, 412–415.

Davenport, D. (1950). Studies in the physiology of commensalism. 1. The polynoid genus *Arctonoë*. *Biol. Bull. mar. biol. Lab. Woods Hole* **98**, 81–93.

Davenport, D. (1953). Studies on the physiology of commensalism—III. The polynoid genera *Acholoë*, *Gattyana* and *Lepidasthenia*. *J. mar. biol. Ass. U.K.* **32**, 161–173.

Davenport, D. and Hickok, J. F. (1951). Studies in the physiology of commensalism. 2. The polynoid genera *Actinoë* and *Halosydna*. *Biol. Bull. mar. biol. Lab. Woods Hole* **100**, 71–83.

Davenport, D., Camougis, G. and Hickok, J. F. (1960). Analyses of the behaviour of commensals in host-factor. I. A hesionid polychaete and a pinnotherid crab. *Anim. Behav.* **8**, 209–218.

Davenport, D., Ross, D. M. and Sutton, L. (1961). Remote control of nematocyst discharge in the attachment of *Calliactis parasitica* to the shells of hermit crabs. *Vie Milieu* **12**, 197–209.

Dimock, R. V. and Davenport, D. (1971). Behavioral specificity and the induction of host recognition in a symbiotic polychaete. *Biol. Bull. mar. biol. Lab. Woods Hole* **141**, 472–484.

Edmunds, M. (1968). Acid secretion in some species of Doridacea Molusca), Nudibranchiata). *Proc. malac. Soc. Lond.* **38**, 121–133.

Edwards, E. (1966). Mating behaviour in the European edible crab (*Cancer pagurus* L.). *Crustaceana* **10**, 23–30.

Entwistle, N. and Pratt, A. D. (1968). 23ξHydroxy-lanosterol: a new triterpene fungal metabolite of the basidiomycete *Scleroderma aurantum* Pers. *Tetrahedron* **24**, 3949–3953.

Erspamer, V. and Glässer, A. (1957). The pharmacological actions of murexine (urocanylocholine). *Br. J. Pharmac. Chemother.* **12**, 176–184.

Etges, F. J. and Decker, C. L. (1963). Chemosensitivity of the miracidium of *Schistosoma mansoni* to *Australorbis glabratus* and other snails. *J. Parasit.* **49**, 114–116.

Fänge, R. (1963). Toxic factors in starfishes. *Sarsia* **10**, 19–21.

Feder, H. M. (1963). Gastropod defensive responses and their effectiveness in reducing predation by starfishes. *Ecology* **44**, 505–512.

Feder, H. M. (1967). Organisms responsive to predatory sea-stars. *Sarsia* **29**, 371–394.

Feder, H. M. and Arvidsson, J. (1967). Studies on a sea-star (*Marthasterias glacialis*) substance responsible for avoidance reactions in the whelk *Buccinum undatum*. *Ark. Zool.* **19**, 369–379.

Feder, H. M. and Lasker, R. (1964). Partial purification of a substance from starfish tube feet which elicits escape responses in gastropod molluscs. *Life Sci.* **3**, 1046–1051.

Fenchel, T. (1965). Feeding biology of the sea-star *Luidia sarsi* Düben and Koren. *Ophelia* **2**, 223–236.

Fischer, P. (1869). Note sur la natation de *Pecten maximus*. *J. Conch., Paris* **17**, 121–123.

Fischer, R. (1971). Gustatory, behavioral and pharmacological manifestations of chemoreception in man. *In* "Gustation and Olfaction; Food Science and Technology" (G. Ohloff and A. F. Thomas, eds.), pp. 187–235. Academic Press, London and New York.

Flury, F. (1915). Uber das Aplysiengift. *Arch. exp. Path. Pharmak.* **79**, 250–263.

Fontaine, A. R. (1964). The integumentary mucous secretions of the ophiuroid *Ophiocomina nigra*. *J. mar. biol. Ass. U.K.* **44**, 145–162.

Forster, G. R. (1951). The biology of the common prawn, *Leander serratus* Pennant. *J. mar. biol. Ass. U.K.* **30**, 333–360.

Fox, H. M. (1924). The spawning of echinoids. *Proc. Camb. phil. Soc. biol. Sci.* **1**, 71–74.

Friess, S. L. (1972). Mode of action of marine saponins on neuromuscular tissues. *Fedn. Proc. Fedn. Am. Socs. exp. Biol.* **31**, 1146–1149.

Gage, J. (1966a). Observations on the bivalves *Montacuta substriata* and *M. ferruginosa*, "commensals" with spatangoids. *J. mar. biol. Ass. U.K.* **46**, 49–70.

Gage, J. (1966b). Experiments with the behaviour of the bivalves *Montacuta substriata* and *M. ferruginosa*, "commensals" with spatangoids. *J. mar. biol. Ass. U.K.* **46**, 71–88.

Galtsoff, P. S. (1938). Physiology of reproduction of *Ostrea virginica*. II.

Stimulation of spawning in the female oyster. *Biol. Bull. mar. biol. Lab. Woods Hole* **75**, 286–307.

Galtsoff, P. S. (1940). Physiology of reproduction of *Ostrea virginica*. III. Stimulation of spawning in the male oyster. *Biol. Bull. mar. biol. Lab. Woods Hole* **78**, 117–135.

Gemmill, J. (1914). The development and certain points in the adult structure of *Asterias rubens*. *Phil. Trans. R. Soc.* **B 205**, 213–294.

Goad, L. J., Rubenstein, I. and Smith, A. G. (1972). The sterols of echinoderms. *Proc. R. Soc.* **B 180**, 223–246.

Gonor, J. J. (1965). Predator-prey relations between two marine prosobranch gastropods. *Veliger* **7**, 228–232.

Gonor, J. J. (1966). Escape responses of North Borneo strombid gastropods elicited by the predatory prosobranchs *Aulica vespertilio* and *Conus marmoreus*. *Veliger* **8**, 226–230.

Gore, R. H. (1966). Observations on the escape response in *Nassarius vibex* (Say) (Mollusca: Gastropoda). *Bull. mar. Sci.* **16**, 423–434.

Haldane, J. B. S. (1955). Animal communication and the origin of human language. *Sci. Prog., Lond.* **43**, 385–401.

Hashimoto, Y. and Kamiya, H. (1969). Occurrence of a toxic substance in the skin of a sea bass, *Pogonoperca punctata*. *Toxicon* **7**, 65–70.

Hashimoto, Y. and Oshima, Y. (1972). Separation of grammistins A, B and C from a soapfish *Pogonoperca punctata*. *Toxicon* **10**, 279–284.

Hickok, J. and Davenport, D. (1957). Further studies in the behavior of commensal polychaetes. *Biol. Bull. mar. biol. Lab. Woods Hole* **113**, 397–406.

Hodgson, E. S., Mathewson, R. F. and Gilbert, P. W. (1967). Electroencephalographic studies of chemoreception in sharks. *In* "Sharks, Skates and Rays" (P. W. Gilbert, R. F. Mathewson and D. P. Rall, eds.), pp. 491–501. John Hopkins Press, Baltimore, Maryland.

Ikegami, S., Kamiya, Y. and Tamura, S. (1972a) A new sterol from asterosaponins A and B. *Tetrahedron Letters*, 1601–1604.

Ikegami, S., Kamiya, Y. and Tamura, S. (1972b). A novel steroid, 3β, 6α, 23ξ-trihydroxy-5α-cholest-9(11)-ene, from asterosaponins. *Tetrahedron Letters*, 3725–3728.

Ikegami, S., Kamiya, Y. and Tamura, S. (1972c). Isolation and identification of spawning inhibitors from ovaries of the starfish *Asterias amurensis*. *Agric. biol. Chem.* **36**, 1087–1089.

Jacobson, S. M. and Boylan, D. B. (1973). Effect of seawater soluble fraction of kerosene on chemotaxis in a marine snail, *Nassarius obsoletus*. *Nature, Lond.* **241**, 213–215.

Jensen, M. (1966). The response of two sea-urchins to the sea-star *Marthasterias glacialis* (L.) and other stimuli. *Ophelia* **3**, 209–219.

Johannes, R. E. (1963). A poison-secreting nudibranch (Mollusca, Opisthobranchia). *Veliger* **5**, 104–105.

Kanatani, H. (1967). Mechanism of starfish spawning with special reference to gonad-stimulating substance (GSS) and meiosis-inducing substance (MIS) of gonad. *Jap. J. exp. Morph.* **21**, 61–78.

Kanatani, H., and Shirai, H. (1968). Does GSS or MIS act as a pheromone?—Problems concerning the participation of a pheromone in starfish spawning. *Zool. Mag., Tokyo* **77**, 207–212.

Kanatani, H., Shirai, H., Nakanishi, K. and Kurokawa, T. (1969). Isolation and identification of MIS in starfish *Asterias amurensis*. *Nature, Lond.* **221**, 273–274.

Karlson, P. and Lüscher, M. (1959). "Pheromones": a new term for a class of biologically active substances. *Nature, Lond.* **183**, 55–56.

Kearn, G. C. (1967). Experiments on host-finding and host-specificity in monogenean skin parasite *Entobdella soleae*. *Parasitology* **57**, 585–605.

Kittredge, J. S. (1971). Marine-fouling organisms: natural attractants and repellants. *U.S. Nat. Tech. Inform. Serv. A.D. Rep.* **1971**, No. 734106.

Kittredge, J. S. and Takahashi, F. T. (1972). The evolution of sex pheromone communication in the Arthropoda. *J. theor. Biol.* **35**, 467–471.

Kittredge, J. S., Terry, M. and Takahashi, F. T. (1971). Sex pheromone activity of the molting hormone, crustecdysone, on male crabs (*Pachygrapsus crassipes*, *Cancer antennarius* and *C. anthonyi*). *Fishery Bull. Fish Wildl. Serv. U.S.* **69**, 337–343.

Laurenson, D. F. (1970). Behavioural and physiological studies of the escape response elicited in members of the Trochidae by thaid and starfish predators. Ph.D. Thesis, University of Auckland.

Lecomte, J. (1953). Reactions de fuite des *Pectens* en presence des asterides. *Vie Milieu* **3**, 57–60.

Machlis, L., Nutting, W. H. and Rapoport, H. (1968). The structure of sirenin. *J. Am. chem. Soc.* **90**, 1674–1675.

Mackie, A. M. (1970) Avoidance reactions of marine invertebrates to either steroid glycosides of starfish or synthetic surface-active agents. *J. exp. mar. Biol. Ecol.* **5**, 63–69.

Mackie, A. M. and Shelton, R. G. J. (1972). A whole-animal bioassay for the determination of the food attractants of the lobster *Homarus gammarus*. *Mar. Biol.* **14**, 217–221.

Mackie, A. M. and Turner, A. B. (1970). Partial characterization of a biologically active steroid glycoside isolated from the starfish *Marthasterias glacialis*. *Biochem. J.* **117**, 543–550.

Mackie, A. M., Lasker, R. and Grant, P. T. (1968). Avoidance reactions of a mollusc *Buccinum undatum* to saponin-like surface-active substances in extracts of the starfish *Asterias rubens* and *Marthasterias glacialis*. *Comp. Biochem. Physiol.* **26**, 415–428.

McInnis, A. J. (1965). Responses of *Schistosoma mansoni* miracidia to chemical attractants. *J. Parasit.* **51**, 731–746.

McLeese, D. W. (1970). Detection of dissolved substances by the American lobster (*Homarus americanus*) and olfactory attraction between lobsters. *J. Fish. Res. Bd Can.* **27**, 1371–1378.

Maretzki, A. and del Castillo, J. (1967). A toxin secreted by the soapfish *Rypticus saponaceus*. A preliminary report. *Toxicon* **4**, 245–250.

Margolin, A. S. (1964a). A running response of *Acmaea* to sea-stars. *Ecology* **45**, 191–193.

Margolin, A. S. (1964b). The mantle response of *Diodora aspera*. *Anim. Behav.* **12**, 187–194.

Mauzey, K. P., Birkeland, C. and Dayton, P. K. (1968). Feeding behavior of asteroids and escape responses of their prey in the Puget Sound region. *Ecology* **49**, 603–619.

Michael, R. P., Keverne, E. B. and Bonsall, R. W. (1971). Pheromones: isolation of male sex attractants from a female primate. *Science, N.Y.* **172**, 964–966.

Miller, R. L. (1963). Chemotaxis in *Campanularia* sperm: species specificity. *Biol. Bull. mar. biol. Lab. Woods Hole* **125**, 386.

Miller, R. L. (1964). Isolation of the chemotactant of *Campanularia* (*Laomedea*) *calceolifera*. *Biol. Bull. mar. biol. Lab. Woods Hole* **127**, 381.

Miller, R. L. (1965). Chemotaxis (?) of coelenterate sperm. *Biol. Bull. mar. biol. Lab. Woods Hole* **128**, 391.

Miller, R. L. and Nelson, L. (1962). Evidence of a chemotactic substance in the female gonangium of *Campanularia*. *Biol. Bull. mar. biol. Lab. Woods Hole* **123**, 477.

Müller, D. G. and Jaenicke, L. (1973). Fucoserraten, the female sex attractant *of Fucus serratus* L. (Phaeophyta). *FEBS Letters* 30, 137–139.

Müller, D. G., Jaenicke, L., Donike, M. and Akintobi, T. (1971). Sex attractant in a brown alga: chemical structure. *Science, N.Y.* **171**, 815–817.

Nigrelli, R. F. (1963). "Metabolites of the Sea". A.I.B.S. B.S.C.S. Pamphlet No. 7. D.C. Heath, Boston, Massachusetts.

Nigrelli, R. F., Stempien, M. F., Ruggieri, G. D., Liguori, V. R. and Cecil, J. T. (1967). Substances of potential biomedical importance from marine organisms. *Fedn. Proc. Fedn. Am. Socs. exp. Biol.* **26**, 1197–1205.

Paine, R. T. (1969). The *Pisaster—Tegula* interaction: prey patches, predator food preferences and intertidal community structure. *Ecology* **50**, 950–961.

Passano, L. M. (1957). Prey-predator recognition in the invertebrates. *In* "Recent Advances in Invertebrate Physiology" (B. Y. Scheer, ed.), pp. 37–47. University of Oregon, Eugene.

Peters, R. L. (1964). The function of the cephalic tentacles in *Littorina*. *Veliger* **7**, 143–148.

Portmann, J. E. (1972). Toxicity-testing with particular reference to oil-removing materials and heavy metals. *In* "Marine Pollution and Sea Life" (M. Ruivo, ed.), pp. 217–222. Fishing News (Books) Ltd., London.

Quayle, D. B. (1940). Japanese oyster propagation. *Prog. Rep. Pacif. biol. Stn. Nanaimo* **43**, 8–9.

Rio, G. J., Stempien, M. F., Nigrelli, R. F. and Ruggieri, G. D. (1965). Echinoderm toxins—I. Some biochemical and physiological properties of toxins from several species of Asteroidea. *Toxicon* **3**, 147–155.

Robson, E. A. (1961a). Some observations on the swimming behaviour of the anemone *Stomphia coccinea*. *J. exp. Biol.* **38**, 343–363.

Robson, E. A. (1961b). The swimming response and its pacemaker system in the anemone *Stomphia coccinea*. *J. exp. Biol.* **38**, 685–694.

Ross, D. M. (1967). Behavioural and ecological relationships between sea anemones and other invertebrates. *Oceanogr. mar. Biol. A. Rev.* **5**, 291–316.

Ross, D. M. (1971). Protection of hermit crabs (*Dardanus spp.*) from octopus by commensal anemones (*Calliactis spp.*). *Nature, Lond.* **230**, 401–402.

Ross, D. M. and Sutton, L. (1961a). The response of the sea anemone *Calliactis parasitica* to shells of the hermit crab *Pagurus bernhardus*. *Proc. R. Soc.* **B 155**, 266–281.

Ross, D. M. and Sutton, L. (1961b). The association between the hermit crab *Dardanus arrosor* (Herbst) and the sea anemone *Calliactis parasitica* (Couch). *Proc. R. Soc.* **B 155**, 282–291.

Ross, D. M. and Sutton, L. (1964). Inhibition of swimming response by food and of nematocyst discharge during swimming in sea anemone *Stomphia coccinea*. *J. exp. Biol.* **41**, 751–757.

Ross, D. M. and Sutton, L. (1967). Swimming sea anemones of Puget Sound: swimming of *Actinostola new species* in response to *Stomphia coccinea*. *Science, N.Y.* **155**, 1419–1421.

Rothschild, the Lord (1956). "Fertilization". Methuen and Co. Ltd., London.

Ryan, E. P. (1966). Pheromone: evidence in a decapod crustacean. *Science, N.Y.* **151**, 340–341.

Sastry, A. N. and Menzel, R. W. (1962). Influence of hosts on the behaviour of the commensal crab *Pinnotheres maculatus* Say. *Biol. Bull. mar. biol. Lab. Woods Hole* **123**, 388–395.

Sheikh, Y. M., Tursch, B. M. and Djerassi, C. (1972). 5α-Pregn-9(11)-ene-3β, 6α-diol-20-one and 5α-cholesta-9(11), 20(22)-diene-3β, 6α-diol-23-one. Two novel steroids from the starfish *Acanthaster planci*. *J. Am. chem. Soc.* **94**, 3278–3280.

Smith, A. G. and Goad, L. J. (1971a). The metabolism of cholesterol by the echinoderms *Asterias rubens* and *Solaster papposus*. *F.E.B.S. Letters* **12**, 233–235.

Smith, A. G., and Goad, L. J. (1971b). Sterol biosynthesis in the starfish *Asterias rubens* and *Henricia sanguinolenta*. *Biochem. J.* **123**, 671–673.

Snyder, N. and Snyder, H. (1970). Alarm response of *Diadema antillarum*. *Science, N.Y.* **168**, 276–278.

Stamm, R. A. (1968). Defense against predators by *Lobiger serradifalci* (Calcara) 1840 and *Oxynoe olivacea* Rafinesque 1819 (Mollusca, Gastropoda, Opisthobranchia). *Revue suisse Zool.* **75**, 661–665.

Stewart, W. C. (1970). Nature of the attractant emitted by the asteroid host of the commensal polychaete, *Ophiodromus pugettensis*. Ph.D. Thesis, University of California.

Szal, R. (1971). "New" sense organ of primitive gastropods. *Nature, Lond.* **229**, 490–492.

Thomas, G. E. and Gruffydd, Ll.D. (1971). The types of escape reactions elicited in the scallop *Pecten maximus* by selected sea-star species. *Mar. Biol.* **10**, 87–93.

Turner, A. B., Smith, D. S. H. and Mackie, A. M. (1971). Characterization of the principal steroidal saponins of the starfish *Marthasterias glacialis*: structures of the aglycones. *Nature, Lond.* **233**, 209–210.

Vasserot, J. (1964). "Defense passive" de l'oursin *Arbicia bixula* contre *Marthasterias glacialis*. *Vie Milieu* **Suppl. 17**, 173–176.

Ward, J. (1965). An investigation on the swimming reactions of the anemone *Stomphia coccinea*. I. Partial isolation of a reacting substance from the asteroid *Dermasterias imbricata*. *J. exp. Zool.* **158**, 357–364.

Went, F. W. (1970). Plants and the chemical environment. *In* "Chemical Ecology" (E. Sondheimer and J. B. Simeone, eds.), pp. 71–82. Academic Press, London and New York.

Whittaker, R. H. and Feeny, P. P. (1971). Allelochemics: chemical interactions between species. *Science, N.Y.* **171**, 757–770.

Whittaker, V. P. (1960). Pharmacologically active choline esters in marine gastropods. *Ann. N.Y. Acad. Sci.* **90**, 695–705.

Willows, A. O. D. and Hoyle, G. (1969). Neuronal network triggering a fixed action pattern. *Science, N.Y.* **166**, 1549–1551.

Wilson, E. O. (1968). Chemical systems. *In* "Animal Communication" (T. Sebeok, ed.). Indiana University Press, Bloomington, Indiana.

Wilson, E. O. and Bossert, W. H. (1963). Chemical communication among animals. *Recent Prog. Horm. Res.* **19**, 672–710.

Winkler, L. R. (1961). Preliminary tests of the toxin extracted from Californian sea hares of the genus *Aplysia*. *Pacif. Sci.* **15**, 211–214.

Wood, L. (1968). Physiological and ecological aspects of prey selection by the marine gastropod *Urosalpinx cineria*. *Malacologia* **6**, 267–320.

Wright, H. O. (1973). Effect of commensal hydroids on hermit crab competition in the littoral zone of Texas. *Nature, Lond.* **241**, 139–140.

Yarnall, J. L. (1964). The response of *Tegula funebralis* to starfish and predatory snails (Mollusca, Gastropoda). *Veliger* **6**, (suppl.), 56–58.

Yasumoto, T., Tanaka, M. and Hashimoto, Y. (1966). Distribution of saponins in echinoderms. *Bull. Jap. Soc. scient. Fish.* **32**, 673–676.

Yentsch, C. S. and Pierce, D. C. (1955). "Swimming" anemone from Puget Sound. *Science, N.Y.* **122**, 1231–1233.

4. Chemoreception in Aquatic Invertebrates with Special Emphasis on the Feeding Behavior of Coelenterates

HOWARD M. LENHOFF

Department of Developmental and Cell Biology, University of California, Irvine, California 92664,

and

K. JUNE LINDSTEDT

Department of Biology, California Lutheran College, Thousand Oaks, California 91360, U.S.A.

I. Introduction

Aquatic invertebrates have proven to be particularly favourable material for chemoreception studies, from the behavioral to the molecular. There seems to be a wide number of instances, especially among the more primitive phyla, in which the chemical senses are the primary ones used for communication and coordination. Perhaps chemical senses evolve more readily, requiring as a minimum no more than a specific protein receptor on a cell membrane. Or, perhaps the aquatic environment provides these relatively simple organisms with a medium well suited for chemoreceptor communication.

From the wide number of chemoreceptor phenomena observed among aquatic invertebrates, we find the chemically controlled feeding behavior of coelenterates especially useful for the experimental study of the most common features of cell chemoreceptors—that is, the combination of a specific activator with its complementary receptor, and the initial events triggered by that combination.

In this review, we discuss chemoreception in aquatic invertebrates very much in the historical way the field has progressed. First, we present a partial survey of a wide number of instances in which behavioral—physiological phenomena are controlled by chemoreception. Second, is a detailed discussion of mechanisms of chemoreception as it occurs in the feeding behavior of hydra and some other coelenterates. And third, we consider the evolution of chemoreceptors and the universality of their action.

II. Survey

In this survey we discuss a number of interesting situations among aquatic organisms in which chemical receptors are thought to be involved. In addition to marine invertebrates, we also include a discussion of bacteria, white blood cells, and freshwater organisms because they share a common simplicity in their chemoreceptor phenomena. We found making our survey exciting, not so much because of the answers discovered, but because of the numerous possibilities that seem to be begging for further experimentation.

Throughout this survey we characterize different chemical effectors involved in feeding according to the recent modifications of Beck's (1965) terminology. These definitions, as put forth by Lindstedt, (1971a), are as follows:

1. Attractant—a stimulus to which an animal responds by orienting toward or becoming receptive to the apparent source. May operate over long distances.

2. Arrestant—a stimulus that causes an animal to cease locomotion when in close contact with the apparent source.

3. Repellent—a stimulus that causes the animal to orient away from or become non-receptive to the apparent source.
4. Incitant—a stimulus that evokes initiation of feeding (tasting).
5. Suppressant—a stimulus that inhibits or prevents initiation of feeding.
6. Stimulant—a stimulus that promotes ingestion and continuation of feeding.
7. Deterrent—a stimulus that prevents continuation of feeding or hastens termination of feeding.

A. Bacteria

Chemotaxis has been observed in *Escherichia coli*. Adler (1966, 1969) reported migration of *E. coli* along a concentration gradient in response to certain of their metabolites. Positive responses were observed to oxygen, glucose, galactose, serine and aspartic acid. Adler (1969) concluded that these substances do not have to be extensively metabolized in order for them to attract bacteria. The systems that detect the chemicals can rightly be termed chemoreceptors since there are mutants of *E. coli* which, while they are not attracted to a particular compound, still metabolize it normally; hence, these mutants are regarded as lacking specific chemoreceptors. Adler (1969) reported at least five different chemoreceptors in *E. coli*. These bacteria respond to attractants in concentrations as low as 10^{-7} or 10^{-8}M. Because of the relative ease with which it is possible to obtain chemoreceptor mutants in bacteria, Adler's system gives much promise for identifying the chemoreceptor substance(s) and uncovering a mechanism of chemoreceptor action.

B. Protozoa and Single Cells

1. *Ciliates*

Chemotactic orientation has been observed in some predatory ciliates. Both *Spathidium spathula* (Woodruff and Spencer, 1922) and *Lionotus lamella* (Dragesco, 1962) orient toward prey ciliates. This response is most pronounced when prey animals are broken up and leak fluids into the medium. Mugard (1949) found that the histophage ciliate, *Orphyoglena* sp. is attracted to an alcohol soluble, thermostable compound isolated from its prey. Protein precipitates, ribonucleic acid (RNA), deoxyribonucleic acid (DNA), adenosine triphosphate (ATP) and peptones all induce positive chemotaxis in this ciliate; i.e., they serve as attractants. Smaller molecules such as amino acids, purines, pyrimidines, nucleosides and nucleotides other than ATP did not attract *Orphyoglena*.

Members of another ciliate genus, *Nassula*, feed exclusively on blue-green algae. Poilvert (1959) has shown that certain compounds isolated from alcohol extracts of cyanophytes serve as attractants for *N. aurea*, *N. picta* and *N. ornata*.

Fauré-Fremiet (1967) observed chemotaxis in members of the following histophage genera: *Coleps*, *Plagiocampa*, and *Proroden* among the Gymnostomatidia, *Philaster*, *Porpostoma*, *Deltophylum* and *Tetrahymena* among the Hymenostomatidia. These ciliates begin to cruise around a fragment of animal tissue or a small, injured organism which they approach progressively. When contact is made they attack the prey.

2. *Sarcodinia and White Blood Cells*

Consideration of chemoreception in protozoa and such single cells as white blood cells should not be restricted to chemotaxis. In fact, it is not inconceivable that chemoreception may have first evolved in single cells along with another series of phenomena—i.e., pinocytosis ("cell drinking") and phagocytosis ("cell eating"). It is well known that certain classes of substances induce these processes.

Most research in this field has involved either *Amoeba proteus* or mouse macrophages. Work on amoeba (Nachmias (1968); Chapman-Andreson (1962); Josefsson (1968); Cooper (1968)) in general shows that a wide spectrum of substances induce pinocytosis. Most effective were cationic substances. These substances include some proteins, basic dyes, and simple salts. Mouse macrophages (Cohn and Parks, 1967a, b) on the other hand, seem more responsive to anionic substances. Interestingly, the amino acids, glutamic and aspartic acids, and adenine nucleotides (and even adenine nucleosides in macrophages) were especially effective in both the amoeba and macrophages.

C. Platyhelminthes

1. *Turbellaria*

The planarian, *Dugesia dorotocephala*, shows positive chemotaxis toward two amino acids, glutamine and lysine, at concentrations of 5×10^{-6}M. Analogs of these compounds, such as aspartic acid, asparagine, α-keto glutaric acid and glutamic acid do not induce positive responses. Chemotaxis was observed to some other amino acids (alanine, cystine, proline, arginine, ornithine) but only at concentrations of 10^{-4}M or higher (Coward and Johannes, 1969).

2. *Trematoda*

The existence of a chemoreceptor in miricidia of the trematode *Schistosoma mansoni* has been demonstrated by Chernin (1970). He finds that the miricidia orient towards low concentrations of a substance emitted by snails, their next host. This "host factor", though not identified, is thermostable non-oxidizable and has a molecular weight of less than 500 (Chernin, 1970).

D. Rhyncocoela

Jennings (1960) reported that specimens of *Lineus ruber* could detect the presence of food at greater distances than would be possible if eyespots were the only receptor organs. He describes the ciliated cephalic grooves as the site of the chemoreceptors and shows them to respond to low concentrations of chemical stimuli.

E. Annelida

The onuphid polychaete, *Diopatra cuprea*, continually tests the medium for the presence of food stimuli by drawing water into its tube. When threshold stimuli are encountered the worm gives a typical feeding response. Mangum and Cox (1966) and Graham and Mangum (1969) reported that several substances isolated from extracts of the soft shelled clam, *Mya arenaria*, induce the response. These are serine, tyrosine, valine and phenylalanine plus and unidentified compound of larger molecular weight (about 20,000–40,000). They suggest that the large molecule might be a glycoprotein or a polypeptide.

Case (1962) and Case and Gwilliam (1963) reported the errant polychaete, *Nereis virens*, able to detect trace amounts of amino acids and alcohols. The type of response (positive or negative) depends on the concentration of the chemical stimulus. Case and Gwilliam (1963) also reported that *Chaetopterus* responds positively to some amino acids and negatively to others. A positive response consisted of an increase in pumping rate, whereas a response was called negative when the flow of water through the tube was abruptly reversed. The amino acids tested were not named.

F. Mollusca

Numerous phenomena controlled by chemoreception have been demonstrated among molluscs, especially among the gastropods. These phenomena include food detection and ingestion, escape behavior, sex determination and attraction, induction of metamorphosis, and settling behavior.

Gastropoda

Adult *Nassarius obsoletus* rely on chemoreception in feeding. Carr (1967a, b) reported that these gastropods responded to substances diffusing from dead animals by repeatedly extending the proboscis. Lactic acid (5×10^{-4}M) and glycine (10^{-3}M) activated the response. Carr (1967b) suggested that the proboscis response is probably due to the combined effect of a number of compounds because the response of *Nassarius* to crude animal extracts is greater than its response to any one compound.

More recently Gurin and Carr (1971) have shown that dilute human plasma or fluids from the oyster *Crassostrea virginica* were remarkably active in stimulating feeding in *Nassarius*. The active components respectively seem to be human serum albumin (at 2×10^{-9}M) and an oyster glycoprotein (at 2×10^{-10}M). The oyster protein (M.W. 120,000) accounts for 90% of the activity in the original fluid.

Some members of the gastropod genus *Conus* rely exclusively on live fish for food (Kohn, 1956). These animals (*Conus striatus* and *C. cactus*) become active within a few seconds after a fish is introduced into the aquarium. They exhibit the same reaction when presented with water in which fish have been kept. Thus the initial stimulus which attracts *Conus* to its prey appears to be a chemical one. It is of interest that no response is observed to a recently killed fish.

The Hawaiian sea-hare, *Aplysia juliana*, feeds exclusively on algae of one species, *Ulva lactuca*. Frings and Frings (1965) reported that when this algae was introduced into holding tanks where *Aplysia* were housed the animals immediately oriented toward the algae. The same response occurred when animals were stimulated by seawater in which *Ulva* was kept for a few minutes. When the water was dropped directly on the mouth, *Aplysia* initiated a normal feeding behavior (Frings and Frings, 1965).

Jahan-Parwar (1969) presented electrophysiological evidence that chemoreceptor elements are located in the tentacles in *Aplysia california*. The application of sea weed extract to tentacles caused marked changes in the spontaneous activity of certain cells in the cerebral ganglion; seawater without the extract had no effect, Other test substances, such as amyl acetate and coumarin (both at 10^{-4}M) were also effective.

There is increasing evidence that the osphradium is an important organ of chemoreception in gastropods. Brown and Noble (1960) found that the gastropod, *Bullia*, emerges from its burrow in the sand when a food stimulus is presented. If the osphradial nerve is cut this behavior is not observed. Bailey and Laverack (1963) recorded electrophysiological responses of the osphradial nerve of *Bullia* on exposure to various chemical stimuli. No responses were recorded to particles of various sizes or to changes in pH. Two dicarboxylic amino acids, aspartate and glutamate, as well as *Mytelus* tissue fluids caused the most intensive responses in *Bullia* (Bailey and Laverack, 1963).

Wood (1968) examined the behavioral, physiological and ecological relationships between molluscan predator *Urosalpinx cinerea* and its prey on the east coast of the U.S. and introduced the concept of ingestive conditioning where the predator's tendency to respond to the odour of a given prey is increased after it has ingested living tissue from that species.

G. Arthropoda

1. *Merostomata*

The horseshoe crab, *Limulus*, has contact chemoreceptors located on the spines of coxal gnathobases of walking legs, on the chelae of prosomal appendages (including chelicerae), and on the spines of the chilaria. Barber and Hayes (1963) reported that these receptors gave strongest electrophysiological responses to aqueous extracts of marine clams, yeast and beef. Some amino acids induced activity in the receptors but only at fairly high concentrations (0·5 M). No single amino acid elicited as strong a response as did the clam extract (Barber and Hayes, 1963).

2. *Crustacea*

Case *et al.* (1960) and Case and Gwilliam (1963) reported that members of seven crab genera respond to amino acids in trace amounts: *Cancer callinectes*, *Carcinides*, *Libinia*, *Pachygrapsus* and *Pagurus*. Case (1964) showed that there are chemoreceptors present on the distal leg segments of *Cancer antennarius* and *C. productus* which are particularly responsive to α-amino-*N*-butyric acid, taurine, glutamic acid, and serine. Laverack (1963), on the other hand, found that two other crabs, *Carcinus* and *Portunus*, and the lobster, *Homarus*, show no sensitivity to amino acids. The dactyl (olfactory) chemoreceptors of these animals respond to low concentrations of two compounds widely distributed in animal tissues, trimethylamine oxide (TMO) and betaine. These compounds probably serve as feeding attractants in these crustaceans, and the $(CH_3)_3$-N moiety is most likely the active part of the molecule since it is common to both compounds. Hodgson (1958) reported that TMO, betaine and glycine are feeding incitants in the crayfish *Cambarus* and the spiny lobster *Panulirus*. Both Laverack (1963) and Hodgson (1958) presented behavioral as well as electrophysiological evidence to support their findings.

A spiny lobster, *Panulirus argus*, has chemoreceptors on its appendages and dactyls which are sensitive to low concentrations of glutamic acid and short-chain tertiary amines. Both behavioral and electrophysiological data suggest that these compounds serve as feeding activators (Levandowsky and Hodgson, 1965).

McLeese (1970) tested various amino acids and mixtures for attractiveness to the lobster *Homarus americanus* and found that certain mixtures had greater attractiveness than the constituent amino acids. However, freshly prepared extracts of cod, shrimp or lobster were more attractive than any of these compounds or mixtures.

Experiments conducted with a simple 2-choice maze showed that an entirely synthetic mixture of chemicals, based on the composition of the short-necked clam *Tapes japonica* (Hashimoto *et al.*, 1968), was highly attractive to the shore crab *Carcinus maenas* (Shelton and Mackie, 1971). The attractiveness of this mixture was greater than that of any of its major components. Similarly, an extract of the squid *Loligo vulgaris* was attractive to the lobster *Homarus gammarus* and ion exchange chromatography showed that a mixture of compounds was necessary for this attraction (Mackie and Shelton, 1972). Experiments with synthetic mixtures, based on the composition of squid extract, have confirmed that the mixture was more attractive than the individual components (Mackie, 1973).

The feeding responses in *Lepas* and other cirripedes was stimulated when solutions of amino acids and organic or inorganic cations were applied to cirri and mouthparts (Crisp, 1967). Proteins, peptides, sugars, neutral organic molecules and anions had no effect. *Lepas* was most responsive to glutamic acid, proline, taurine and betaine (at concentrations of 10^{-4}M or below).

The cumacean *Cumella vulgaris* selects its substrate depending upon the grain size and the amount of organic material present (Wieser, 1956). Thus, perhaps chemical as well as physical stimuli influence selection in this species.

H. Echinodermata

1. *Asteroidea*

Chemical control of feeding behavior has been observed in at least two starfishes. Amino acids and some polypeptides induce feeding responses in *Asterias rubens* whereas carbohydrates induce negative responses (Carthy, 1958). Araki (1965) found that protein and some protein derivatives serve as feeding incitants in *Patiria miniata*. Protein hydrolysates, amino acids and carbohydrates are effective to a much lesser extent.

Castilla and Crisp (1970) reported that *Asterias rubens* are attracted by water which has passed over living mussels (*Mytilus edulis*) and barnacles, their normal prey in the Menai Strait, but not by waterborne odors of oysters. If the starfish were conditioned to eat oysters for a month, they were then attracted by live oysters, another example of ingestive conditioning (Castilla, 1972a). The starfish were, however, repelled by mussels which had been opened by cutting the adductor muscle and by extracts of *Mytilus*, L-glutamic acid and a mixture of amino acids (Castilla, 1972b). In contrast, Zafiriou *et al* (1972) reported that *Asterias vulgaris* was attracted by intact oysters and oyster tissue and also by extracts of four shellfish species.

III. Chemical Control of Feeding in Coelenterates—on the Mechanism of Action of the Glutathione Receptor in Hydra

A. Introduction

Perhaps the biggest boost to research on chemoreception in coelenterates came from Loomis' landmark experiment (1955) showing that the ubiquitous tripeptide reduced glutathione specifically activates the "feeding response" of *H. littoralis*. Since this discovery, numerous workers have shown that phases of the feeding behavior of many other coelenterates are under the control of chemicals, and much work has been done on the mechanism of action of the glutathione receptor of hydra.

We divide this section into a discussion of: (1) evidence showing that reduced glutathione is an activator of feeding in *H. littoralis*; (2) other coelenterates whose feeding responses are known to be activated either by glutathione or by other small molecules; (3) a variety of behavioral responses of hydra that are either affected by or elicited by glutathione; and (4) studies on the mechanism of action of the glutathione receptor-effector system of hydra.

B. Reduced Glutathione and the Feeding Response of *H. littoralis*

Feeding in hydra consists of many separate steps: (1) a prey organism that accidentally encounters an outstretched tentacle is captured, wounded, and poisoned through the action of the deadly nematocysts* that line the tentacle; (2) following capture of the prey the tentacles contract toward the mouth and the mouth opens; (3) on contact with the mouth the food is ingested.

The second step, which is the step that we call the feeding response, was first accurately described by Ewer (1947): "the tentacles writhe and twist towards the mouth, while the mouth itself opens widely." It had long been known that extracts of food elicit a feeding response in coelenterates (Nagel, 1892; Beutler, 1924; Abe, 1938; Pantin and Pantin, 1943; and Ewer, 1947). When Loomis (1955) sought to identify the stimulator of feeding in *H. littoralis*, he used Ewer's criteria. By subjecting the food extracts to various treatments, Loomis (1955) showed that the active principal in fresh tissue juice was heat-stable, active in small amounts, and labile both to long exposure to room temperature and to treatment with hydrogen peroxide. Fulton (1963) repeated these experiments on hydra and obtained the same results.

Surmising that the active principal was a small oxidizable molecule, Loomis tested such biological compounds as ascorbic acid, coenzyme A, cysteine, and glutathione. Only reduced glutathione induced a feeding response. Concerned about the possibility that a trace amount of some substance might contaminate his glutathione preparation, Loomis obtained some chemically synthesized glutathione (du Vigneaud and Miller, 1936) free of biological contamination and found that it stimulated a feeding response.

To determine whether hydra responded only to the complete glutathione molecule, and not to any portion of the tripeptide, Loomis (1955) tested compounds related to glutathione. He found γ-glutamylcysteine, cysteinylglycine, glycylcysteine, cysteine, and asparthione to be inactive. It is noteworthy that synthetic asparthione (β-aspartylcysteinylglycine) did not activate a response, and thus provided strong evidence of the specificity of the hydra receptor for glutathione. Asparthione has all the reactive groups of glutathione only in that it lacks one methylene group. Among the various biological systems known to require glutathione, the feeding response of hydra is the only one not also stimulated by asparthione.

The unique specificity of the hydra receptor for glutathione was further documented by studies in which glutathione analogs and related amino acids were used (Cliffe and Waley, 1958; Lenhoff and Bovaird, 1961). Data from all these investigations, summarized in Table 1, established that: (a) The thiol is not required for activation;

* Nematocysts are complex intracellular secretion products diagnostic of the coelenterates, and consisting of a capsule containing an eversible thread continuous with the capsule wall at one end and stoppered there by an operculum.

ophthalmic acid (γ-glutamyl-α-amino-n-butyrylglycine), norophthalmic acid (γ-glutamylalanylglycine), and S-methylglutathione also activated feeding. (b) The hydra recognizes the specific structure of the intact tripeptide backbone of glutathione; this is evident because the

Table 1. Activators and inhibitors of the feeding response. The formula given below represents the basic tripeptide backbone of glutathione and its analogs. For example, when the R of the alanyl component (component B) is -SH, the formula represents reduced glutathione and, hence, is an activator. On the other hand, if R = -S-SG, then the formula represents oxidized glutathione, an inhibitor of the feeding response. (Adapted from Lenhoff 1961a). A, B, and C refer to the three component amino acids of the tripeptide backbone of γ-glutamylalanylglycine.

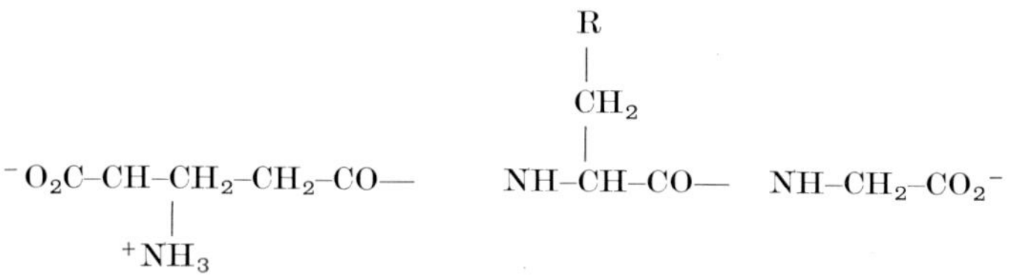

A γ-glutamyl | B alanyl | C glycine

Activators (tripeptide)	Inhibitors	
	Tripeptide	Others
$R = -H$	$R = -SO_2H$	Glutamic acid
$R = -CH_3$	$R = -SO_3H$	Glutamine
$R = -SH$	$R = -S-COCH_3$	Cysteinylglycine
$R = -S-CH_3$	$R = -S(N\text{-ethylsuccinimido})$	
	$R = -S-SG$	
	$R = -SH$; $A = {}^{-}O_2C-CH(NH_3^+)-CH_2CO-$*	

* In this special case A = β-aspartyl rather than γ-glutamyl.

analogs just mentioned activated feeding, and tripeptide analogs with large and charged substituents at the sulfhydryl grouping of glutathione action were inhibitory. (c) The receptor has a high affinity for the glutamyl part of the tripeptide; glutamic acid and glutamine were the only amino acids to show competitive inhibition. (d) The α-amino of glutathione is probably required for the association of glutathione with the receptor; glutamic acid showed competitive inhibition, whereas β-glutamic acid and α-ketoglutaric acid did not.

Knowledge of the inhibitory action of glutamic acid helped prove that reduced glutathione was the substance present in extracts of *Artemia salina* that elicited feeding. Addition of glutamic acid greatly decreased the activity of the extracts, while addition of glutathione to these extracts overcame the inhibition (Lenhoff, 1961a). Glutamic acid was not competing with ophthalmic acid in the extracts because this rare tripeptide, first found in calf lens (Waley, 1958), is not present in *Artemia salina* (Cliffe and Waley, 1958).

The weight of evidence regarding the specificity of the hydra receptor for glutathione was increased when Rushforth *et al.* (1964) showed that the same conformation of glutathione plays another unique role, that of regulating two other receptor-effector systems in hydra, as discussed below.

We have also shown that, under rather special experimental conditions, such nontripeptides as trypsin (Lenhoff and Bovaird, 1960b) and zinc (Lenhoff and Zwisler, 1963) can activate feeding responses. These findings do not contradict the results obtained with glutathione. On the contrary, they may provide us with tools for studying aspects of the mechanism of the glutathione-activated response (Lenhoff, 1965a and 1968a).

C. Feeding Activators Among Other Coelenterates

Loomis, in the last sentence of his 1955 paper states, "The chemical mediator involved (in the feeding reactions of other coelenterates) may consist of glutathione in certain cases as in hydra, or may consist of some other cell constituent that functions in a similar manner." Like so many of his projections, this one also proved true. Since 1963, the feeding responses of about 20 coelenterates have been shown to be chemically controlled, and besides reduced glutathione, 6 specific amino acids have been shown to act as feeding incitants and/or stimulants.

Actually it is not a simple matter to determine the requirement for a feeding activator. Certain "biological" safeguards have to be taken. Of utmost importance are the maintenance of the animal's well-being, and knowledge of its immediate past history. Frequently, animals collected in the field are not healthy when brought into the laboratory. The feeding response may be modified if the animal has been in stagnant water or under crowded conditions. The recency of its last meal will influence its feeding response. Ideally, it would be best to use animals raised in the laboratory: if they are able to increase in mass or reproduce asexually, then they may be presumed to be healthy.

Only after such safeguards have been taken should the feeding peculiarities of the particular species under study be characterized.

At this point, the investigator should check whether phases of the behavior are elicited by nonspecific stimuli. He should also be aware that some coelenterates, especially anemones, may be stimulated to feed on biologically inert objects (cf. Lenhoff, 1965b). To investigate mechanisms and quantitative aspects of coelenterate feeding behavior, an accurate means of quantifying the response is needed.

If the procedures established by Loomis (1955) and by Fulton (1963) are followed, a whole spectrum of amino acids, peptides, and possibly other substances may be found to function as specific activators of feeding in a correspondingly wide range of coelenterates. A big drive to determine the nature of such feeding activators began at a summer program on coelenterate biology at the University of Hawaii (Lenhoff *et al.*, 1971). Perhaps the most interesting features of this newer work are that more than one compound may control feeding, and that feeding may consist of more than one phase, each one of which is controlled by different compounds. These studies, mostly by Lindstedt and Reimer, are described below.

1. *Other Hydra*

Other species of hydra respond to glutathione, but not always in the same manner as *H. littoralis*. For example, *H. pirardi* may respond to reduced glutathione for as long as 100 minutes at 22°C, closing and reopening its mouth many times during that period. In contrast, *H. littoralis* under similar conditions keeps its mouth open continuously for about 30 minutes. *Chlorohydra viridissima*, instead of opening its mouth wide in response to glutathione, as *H. littoralis* does, opens its mouth slightly, sometimes barely detectably, although it can ingest inert material immersed in glutathione (Mariscal, 1971). *H. pseudoligactis*, which responds to free glutathione, is observed occasionally to ingest inert material in the absence of added glutathione. Each species of hydra, therefore, may have its own peculiar feeding behavior. The results from experiments with *H. littoralis* should be used for purposes of comparison and should not be considered representative of all species of hydra.

2. *Other Glutathione Responders*

A dramatic response initiated by glutathione was exhibited by a marine hydrozoan, the siphonophore *Physalia physalis* (Portuguese man-of-war) (Lenhoff and Schneiderman, 1959). The man-of-war is a colonial coelenterate having numerous specialized zooids attached to a float. The function of food capture is carried out by one type of zooid (sometimes 9–12 meters (30–40 feet) long) and the function of ingesting the food is carried out in a coordinated fashion by the hundreds to feeding polyps, called gastrozooids. Lenhoff and Schneiderman (1959) have shown that juice from a fish, or a weak solution (10^{-5}M) of reduced glutathione causes isolated gastrozooids to writhe, twist, and open their mouths. Once the lips of an open mouth touch a solid surface, such as the walls of a container or the side of a captured fish, the lips begin to spread over that surface as though attempting to engulf it. In so doing, the tubelike gastrozooid of about 1 millimeter in diameter becomes transformed into a disk of more than 20 millimeters in diameter. Hundreds of gastrozooids pressing against the surface of a fish in a similar manner can envelop the fish, forming a complete "stomach" around it.

In another siphonophore, *Nanomia cara*, the writhing activities associated with feeding have also been ascribed to the influence of glutathione (Mackie and Boag, 1963). Another hydroid responding positively to glutathione is the calyptoblastic hydroid *Campanularia flexuosa* (Lenhoff and Schneiderman, 1959).

More recently, Reimer (1971a) has shown that a newly described species of *Zoanthus* obtained from Coconut Island, Oahu, Hawaii, responded specifically to reduced glutathione. Unlike hydra, however, *Zoanthus* seemed more responsive to the glycyl moiety of glutathione rather than the glutamyl end.

3. *Proline Responders*

The first truly specific response to a specific chemical other than glutathione was shown by Fulton (1963), using the gymnoblastic colonial hydroid *Cordylophora lacustris*, which lives in brackish water. He identified proline as the activator of the feeding response by extending Loomis' procedures to include chromatographic separation of substances in the tissue extracts. Using proline analogs, he showed that azetidine-2-carboxylic acid, which has a four-membered ring, was almost as active as proline, whereas the six-membered ring analog pipecolic acid had about a tenth of the activity of proline. Through the use of these and other analogs, Fulton (1963) showed that the *Cordylophora* receptor can recognize specifically the imino region of a heterocyclic α-imino acid which is neither substituted nor unsaturated in such a way as to affect the imino acid group.

Another gymnoblastic hydroid, the marine *Pennaria tiarella*, also responded to proline (Pardy and Lenhoff, 1968) at concentrations as low as 10^{-6}M. The proline analog pipecolic acid also elicited a response. No other substances tested, including glutathione, elicted a feeding response in this organism.

The third coelenterate shown to give an unequivocal feeding response to proline was not another gymnoblastic hydroid but the coral *Cyphas-*

trea (Mariscal and Lenhoff, 1968). This scleractinian responded to proline at concentrations of 10^{-7} to 10^{-3}M and to pipecolic acid at concentrations of 10^{-8} to 10^{-3}M. *Cyphastrea* in particular was interesting in that it also gave a feeding response to 10^{-4}M reduced glutathione and to its analog *S*-methylglutathione. Hence, the results with *Cyphastrea* offer the first well-documented case of a coelenterate giving a feeding response to two different types of molecules, proline at low concentration and glutathione at higher concentrations.

A further variation of the involvement of both proline and glutathione was described by Reimer (1971b). She showed that the Hawaiian zoanthid, *Palythoa psammophilia*, responded best to the presence of a mechanical stimulus in a solution containing both proline and reduced glutathione. Either compound alone was also effective, but only at unnaturally high concentrations. The most effective concentrations for these compounds acting synergistically were 5×10^{-5}M proline and 10^{-6}M reduced glutathione (Reimer, 1971b). The evolutionary implications of the findings of Mariscal and Lenhoff (1968) and of Reimer (1971b), are discussed below.

These latter findings point out that we can never say for certain that a coelenterate gives a feeding response to only a single molecule (or its analogs) because it is virtually impossible to test all substances present in tissue extracts. It is possible, however, to show, through use of analogs and through competition experiments (as was done in the case of glutathione and *H. littoralis*) (Lenhoff, 1961a), that the substance under consideration is the major one in the tissue extracts tested that stimulates a feeding response.

4. *Valine Activation*

The feeding response of the Hawaiian swimming actinian *Boloceroides* sp. was recently shown to be controlled by the branched amino acid valine (Lindstedt *et al.*, 1968). Isoleucine, which is basically a valine having an ethyl group instead of one of the branched methyls, is an effective competitive inhibitor. On the other hand, leucine, identical to valine in all respects except that the branch point is separated from the α-carbon by an additional methylene group, is not effective either as an activator or as an inhibitor. Thus, *Boloceroides* can be said to have a receptor specific for an n-α-amino butyric acid with a branch point at the β-carbon.

An important observation made in the *Boloceroides* experiments was that valine did not cause wide mouth opening but, rather, caused the coelenterate to swallow an inert object. Undoubtedly other coelenterates give a distinctive feeding response to a specific chemical activator only in the presence of solid material. In fact, in those reported instances of coelenterates ingesting so-called inert solid objects in the absence of added chemical stimulator—as demonstrated, for example, with *Epiactis prolifera* (Lenhoff, 1965b)—there remains the strong possibility that traces of stimulating compounds were present on the object or in the water.

5. *A Glutamine and a Leucine Responder*

We have been growing in the laboratory a clone of unidentified acontiate pedal-lacerating sea anemone isolated from a floating sargassum weed in Biscayne Bay, Florida in 1964. We showed that this anemone responds to glutamine, though neither glutathione, glutamic acid, nor asparagine would activate a response. Like *Boloceroides*, this anemone did not give an easily recognizable response to its feeding activator unless solid material was also present (Smith and Lenhoff, unpublished).

The sea anemone *Haliplanella* was shown to respond specifically to leucine (Lindstedt, 1971a).

6. *A Biphasic Feeding Response to Two Different Compounds*

Up to this point we have been discussing examples of the complete feeding response being controlled by a specific compound, or two specific compounds, or by a synergistic effect of two specific compounds. Lindstedt (1971b) describes an unusual case in which two phases of the feeding response of the sea anemone *Anthopleura elegantissima*, are controlled by different chemical activators. Asparagine controls the contraction and bending of tentacles which brings food to the mouth; reduced glutathione controls the ingestion of food once it has contacted the mouth. A complete feeding response occurs only when both chemical activators are present.

D. Other Actions of Glutathione on Hydra

Mainly through the efforts of N. Rushforth, glutathione has been shown to influence the physiology and behavior of hydra in five other measurable ways (Fig. 1). It was demonstrated that this tripeptide (1) increased the rate of the tentacle-waving "concerts", discussed below (Rushforth and Hofman, 1972); (2 and 3) inhibited two behavioral contraction responses of hydra which are not part of the feeding response (Rushforth *et al.*, 1964); (4) stimulated changes in bioelectric potential (Rushforth, 1967a, b, 1971; Rushforth and Burke, 1971); and (5) initiated a newly described behavioral response only when another chemoreceptor system was operating at the same time (Blanquet and Lenhoff, 1968).

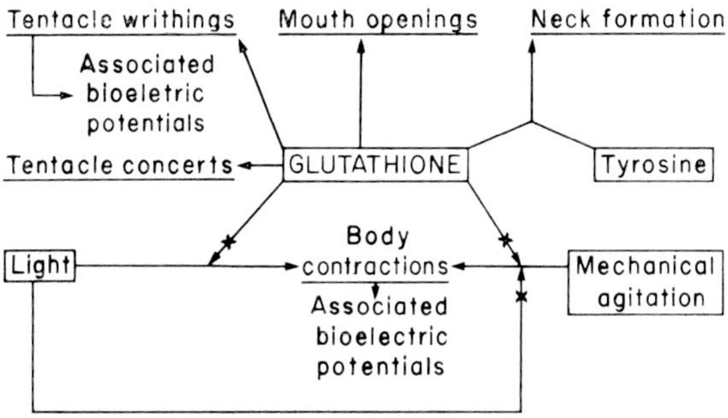

Fig. 1 Integration of receptor-effector systems. This diagram, which summarizes the various measurable effects stimulated by the action of glutathione on hydra, also emphasizes that the output of the glutathione receptor is linked with other receptor-effector systems of the animal. In addition, it points out an action of light in inhibiting the contraction response to mechanical agitation (Rushforth, 1965). An arrow marked with a cross indicates an inhibitory action. (Reprinted from *Science*, with permission.)

1. *Tentacle Concerts*

The flexing motions of the tentacles sweeping inward toward the central vertical axis above the mouth are called tentacle concerts. These concerts occur spontaneously in hydra. Rushforth and Hofman (1972) find that reduced glutathione in a concentration as low as 5×10^{-10}M gives rise to a statistically significant increase in the frequency of tentacle concerts in *H. littoralis*, and that the frequency accelerates with increase in glutathione concentration up to 5×10^{-9}M. At this higher glutathione concentration, tentacle writhing commences.

2. *Glutathione Inhibition of Contraction Responses*

Rushforth (Rushforth *et al.*, 1964; Rushforth, 1965) has been conducting extensive and convincing experiments which show that the contractions of hydra induced either by light or by shaking, as well as the animal's spontaneous contractions, are inhibited while the animals are feeding on *Artemia salina*. Prompted by the experiments on the feeding response, Rushforth tested, first, *Artemia* extracts and then reduced glutathione; both inhibited the contractions. Using his quantitative procedures for measuring the inhibition of the contraction response, Rushforth (1965) showed that the "modes of action of glutathione are similar to those discovered . . . [in] studying the mouth opening response".

In most of his experiments Rushforth used *Hydra pirardi*, European *H. viridis* (symbiotic and aposymbiotic), European stolonizing hydra, and *Chlorohydra viridissima* (Florida strain, 1961; symbiotic and aposymbiotic). Such experiments demonstrate that many species of hydra have a specific glutathione receptor which not only controls but, through its output, affects the contraction responses of the animal.

The mechanism by which glutathione inhibits the contraction response induced by light or mechanical agitation is unknown. Possibly there is a direct means whereby gluthathione turns off the contraction responses, or glutathione may inhibit the response as an indirect consequence of having elicited the contractile events involved in the feeding behavior.

3. *Electrophysiological Correlates of Glutathione-activated Feeding Response*

Extending his research on the contraction responses, Rushforth found both indirect and direct electrophysiological correlates of the glutathione-activated feeding response (Rushforth, 1967a; Rushforth and Burke, 1971). Just as homogenates of *Artemia* or solutions of glutathione inhibited the contraction response of hydra, so they inhibited the production of electrical potentials associated with either the spontaneous contractions of hydra or contractions induced by light. Such potentials were originally described by Passano and McCullough (1962). Furthermore, Rushforth found that the electrical potentials associated with the contraction of isolated tentacles of *H. pseudoligactis* were not produced in the presence of 10^{-5}M reduced glutathione.

More striking is Rushforth's discovery that, when reduced glutathione inhibited the production of potentials associated with tentacle contraction, at the same time it directly initiated potentials associated with the glutathione-induced asymmetric tentacle movements. As the tentacle adapted to glutathione, the frequency with which these potentials were produced decreased, and the spontaneous tentacle contractions with their associated potentials were restored (Rushforth and Burke, 1971). Hence, these experiments not only present the first evidence of direct electrophysiological correlates of glutathione action but also provide strong evidence for the presence of glutathione receptor sites on hydra tentacles.

4. "*Neck*" *Formation*

Neck formation was discovered by Blanquet and Lenhoff (1968) using hydra (mostly *Chlorohydra viridissima* and *H. pirardi*) whose gastrovascular cavity was swollen with fluid and food particles (as observed between 1–6 hours following ingestion of food). Such hydra, when presented with *Artemia* extract or a solution of reduced glutathione, formed a tight constriction in the region just below the hypostome and sometimes extending over the adjacent one-third of the body tube. If, instead of a glutathione solution, the swollen hydra were presented with a live *Artemia* nauplius, the neck constriction formed, the mouth opened, and the hydra swallowed the nauplius. During ingestion the nauplius was carried down through the constriction, apparently by peristaltic contractions, and into the fluids of the swollen gastrovascular cavity. Hence, it would appear that these neck constrictions allow hydra to retain previously ingested food in the gut while swallowing newly captured prey.

Neck formation in *Hydra pirardi* was shown to be caused by a combination of three factors: (a) the presence of glutathione on the exterior of the hydra; (b) distention of the wall of the hydra's body tube; and (c) the presence of tyrosine within the gut. No other natural amino acid, including phenylalanine, could substitute for tyrosine. Analogs of tyrosine having either the α-amino or α-carboxyl blocked were inactive.

From these experimental results we conclude that, in addition to the external gluthathione receptor, hydra has an enteroreceptor specific for tyrosine. The hydroxyl, the α-amino group, and α-carboxyl group must all be present in order for the amino acid to be active (Blanquet and Lenhoff, 1968).

The existence in hydra of two receptor systems that must act in harmony represents, to our knowledge, the first reported instance of two integrated, chemically mediated coordinating systems in the lower invertebrates. This system differs from the asparagine-glutathione system in *Anthopleura* (Lindstedt, 1971b) in which the molecules act in two sequential steps.

E. The Glutathione Receptor

Most information about the glutathione-elicited feeding response in *Hydra littoralis* came from studies in which this system was used as a model for the investigation of the mechanism of activation of a specific chemical receptor site. Such studies required animals that could respond to glutathione in a dependable and quantitative fashion. Of all the coelenterates investigated thus far, *H. littoralis* has proved to be the only one in which reliable quantification of the feeding response was possible.

With *Hydra littoralis* we were able to obtain animals that were genetically alike, in the same stage of development, derived from logarithmically growing cultures, and grown in a defined environment (Lenhoff, 1961b; Loomis and Lenhoff, 1956; Lenhoff and Brown, 1970). Thus, there were always large numbers of animals that could respond to glutathione nearly synchronously. Details of the procedures and of the present assay methods are given elsewhere (Lenhoff, 1961b). Suffice it to say that the major parameter of measurement was the "duration of the feeding response"—that is, the length of time the animal's mouth remained open in the presence of reduced glutathione. This assay, although a measure of behavioral response, is both accurate and objective because the investigator had merely to record the precise times that the hydra's mouth opened and closed. Since 1962 the assay always has been carried out at constant temperature and pH and in a solution of known ionic composition. The experimental hydra were placed directly into a solution of glutathione in order to activate all functioning glutathione receptor-effector systems and, thus, gain further control over the animals. These conditions differ, of course, from those in the pond, where hydra are presented with an oriented gradient of glutathione and of other substances emitted from the prey, in a solution of unknown composition. By controlling our experimental system in the manner described, we procured reproducible results with as few as five animals per measurement.

Studies aiming to uncover the mechanism of action of an excitatory substance can usually take many routes One is the determination of the size and shape of the molecule that is active; this point is discussed above in regard to the excitatory effect of glutathione on hydra. In the rest of this section we deal with three other major questions: (1) What is the nature of the ionic media surrounding the receptor (and effector), and what are the effects of those ions on the response? (2) What type of interaction occurs between activator and receptor? Is it fast or slow? Is the activator metabolized? (3) What are some of the properties of the receptor?

1. *Studies of Inorganic Ions*

Many inorganic ions affect the extent of the response activated by glutathione. Since these ions bathe both the receptor and the ectodermal effector cells, which are involved in part of the contractile processes of the feeding response, it is difficult to determine where and how these ions act. In any case, these studies are important because they define the limits within which the ionic composition can be varied, they reveal previously indiscernible aspects of the physiology of the animal, and they may add to our understanding of the mechanism of activation of the receptor.

Studies of the effect of environmental ions on the physiology of hydra are also of considerable importance in comparisons of work on hydra with work on other organisms. Hydras appear to be unique among members of the animal kingdom in that they are diploblastic, possess essentially no extra-cellular fluids (aside from the contents of the gastrovascular cavity), and live in fresh water. Thus the external medium plays a functional role similar to the *milieu intérieur* of higher forms. It is through this external environment that glutathione passes. Since external environments can be subjected to much more rigorous control than the *milieu intérieur* can be, extensive study of environmental ions (Lenhoff and Bovaird, 1960a; Lenhoff, 1966) should prove valuable.

Without calcium ions, hydra could not respond to reduced glutathione (Lenhoff and Bovaird, 1959). The requirement for calcium was found to be pH-dependent (Lenhoff, unpublished), and a concentration of about 10^{-4}M was necessary for a maximum response. Strontium was the only ion that could substitute for calcium, and even it was much less effective (Lenhoff and Bovaird, 1959). Magnesium ions were not required; in high concentrations they inhibited the responses by competing with calcium ions (Lenhoff, 1961a; Lenhoff and Bovaird, 1959). Sodium likewise competed with calcium, but less effectively than magnesium. The possible sites at which calcium can act in the complete glutathione receptor-effector system would seem to be innumerable.

Potassium ions were found to inhibit the feeding response (Lenhoff, 1966), but, unlike magnesium and sodium ions, they did not act by competing with calcium ions. Concentrations of potassium ions as low as 10^{-4}M could lower the response to glutathione significantly, and this inhibition could be reversed by placing the animals in a potassium-free medium for a few hours (Lenhoff, unpublished). Since potassium ions play an important role in bioelectric potential, it may be that these ions act by affecting the cellular membrane potential of hydra. More recent work using *H. attenuata* shows that the presence of sodium ions in the environment is an absolute requirement for the glutathione-mediated feeding response to occur (Asbill, Danner and Lenhoff, unpublished observations).

The effect of such controllable environmental factors as anions, inhibitors, and temperature is described elsewhere (Lenhoff, 1961a; Lenhoff, 1968a; Lenhoff and Bovaird, 1959).

2. *Glutathione-receptor Interaction*

An indication of the speed at which the equilibrium between glutathione and the receptor was attained was determined by means of some relatively simple experiments. Hydra placed in a glutathione solution would open their mouths within a minute, and they would close their mouths within a minute after the glutathione was removed (Lenhoff, 1961b). These same animals could repeat this opening and closing many times during an hour (Lenhoff, 1961b). Hence we can conclude (a) that glutathione has to be present constantly in the solution, and thus at the receptor site, in order for a response to take place, and (b) that the equilibrium between glutathione and the receptor is rapidly attained.

3. *Is Glutathione Consumed?*

The feeding response induced by glutathione has a finite period. This period is temperature-dependent—for example, about 30 minutes at 22°C (Lenhoff, 1961a). This limitation in duration is not caused by the disappearance of glutathione from the culture solution; that same solution of glutathione (or of the nonoxidizable analogs ophthalmic acid and *S*-methylglutathione), after removal from hydra had made a maximum feeding response, induced a new group of animals to respond (Lenhoff, 1961a). Perhaps the cessation of the response was brought about by consumption of some substance in the receptor-effector system or, alternatively, by the production of an inhibitor. Whatever the correct explanation of this phenomenon, the hydra did not open their mouths in response to glutathione in the hour immediately following the end of a maximum response; in the subsequent 24 hours, however, they gradually regained their full capacity to respond (Lenhoff, 1961a).

The activation of a feeding response does not produce any detectable changes in the structure of reduced glutathione (Lenhoff, 1961a, b). It is possible that changes may have occurred which were too small to be detected by present methods. In any case, it is not necessary to assume that glutathione is altered at all when causing a feeding response. There are known instances, as in enzyme induction by substrate analogs, in which a biological response is initiated by a molecule (noncoenzymic in function) combining with a specific site without that molecule being metabolized.

4. *Behavioral Determination of a Dissociation Constant*

More recent experiments have centered around determination of the dissociation constant between glutathione and the receptor, and around use of the equilibrium data to elucidate the nature of the receptor site in the same way that an enzymologist uses date on K_M (the dissociation of the enzyme-substrate complex) to help determine the active site of an enzyme.

We have reported elsewhere (Lenhoff, 1965a and 1968a) the assumption made in determining the dissociation constant, K_M, between the activator A and the receptor R. The effect of the activation is signified by ε and the maximum effect, by ε_M. The equation derived,

$$\frac{(A)}{\varepsilon} = \frac{1}{\varepsilon_M}(A) + \frac{K_A}{\varepsilon_M}$$

is analogous to the second form of the Lineweaver-Burk (1934) plot, the equation developed by Beidler (1954) for mammalian taste chemoreception and, of course, a form of the Langmuir adsorption isotherm.

As shown in Fig. 2 a plot of $(A)/\varepsilon$ against (A) gives a straight line at most glutathione concentrations. The slope of the line is $1/\varepsilon_M$ and the extrapolated intersect is K_A/ε_M. If the line is further extrapolated, it

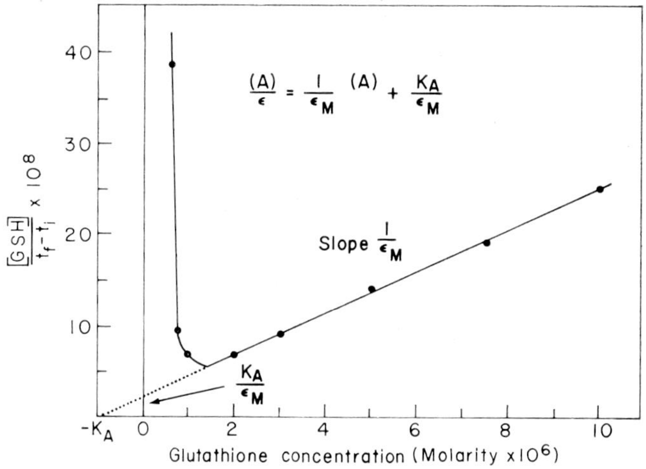

Fig. 2 Plot for determining constants of the combination of glutathione with its receptor. GSH, reduced glutathione. (Reprinted from *American Zoologist*, with permission.)

intersects the abscissa at $-K_A$. Unlike data published according to the Lineweaver-Burk and Beidler plots, at low concentrations of glutathione the curve of Fig. 2 swings asymptotically upward. This upswing is always present and repeatable. It probably represents "below-threshold" activation. That is, at very low concentrations of glutathione the physiological response is not detectable by our behavioral assay; possibly the contractile fibrils involved in the animal's mouth-opening have not yet overcome the actions of those fibers that tend to keep the mouth closed. In the Lineweaver-Burk plot, such an upswing would be observed if the method used to assay the enzyme product were insufficiently sensitive. In the electrophysiological studies of Beidler, a similar upswing might result from plotting data at the "noise" level. The upswing of Fig. 2 gives promise of being a useful quantitative index of threshold. This upswing, however, is never seen at higher concentrations of glutathione; the line at higher concentrations of glutathione is straight and can be used to determine accurately the dissociation constant.

The dissociation constant of about 10^{-6}M, as determined through a plot of this kind, is significant in at least four ways. (a) The smallness of the constant indicates a high affinity of the receptor for glutathione. (b) Concentrations around 10^{-6}M are well within the physiological range to be expected under natural conditions of feeding. (c) This constant provides a means of characterizing the receptor; that is, the glutathione receptor of *H. littoralis* may be said to have a dissociation constant of 10^{-6}M, under the given conditions. The constant is a characteristic of the receptor and remains nearly the same no matter what the nutritional state of the hydra (Lenhoff, 1961a, b). Similarly, experiments in which the buffer anion is varied alter the maximum response, but not the dissociation constant (Lenhoff, 1969). (d) Changes in the K_A with pH can be used to determine the pK's of the ionizable groups of glutathione or at the receptor site which are involved in the combination with glutathione.

5. *The pH Profile of the Glutathione Receptor*

The pK measurements were made by means analogous to those used by enzymologists in determining the pK's of ionizable groups at the active site of enzymes. For our purposes we needed an equilibrium equation, like Dixon's (1953) for enzymes, which would take into account the influence of pH on the dissociation constant. This modified equation (Lenhoff, 1965a; 1968a, b; 1969) involved the assumption that if the activator, receptor site, or activator-receptor complex ionizes, then, in the expression for equilibrium, each component (A, R, AR) equals its concentration multiplied by a term which is a function of pH. For example, if the activator ionized, then the total concentration of free activator, A_t, would be A times the pH function of A, or f_a (pH). The logarithmic form of the equation is:

$$pK_A = pK^\circ_A + \log f_{ar}\,(\text{pH}) - \log f_r\,(\text{pH}) - \log f_a\,(\text{pH}).$$

Here pK_A refers to the negative logarithm of the dissociation constant of AR, while pK°_A is the same constant if none of the components has ionic groups; if none ionizes, then pK_A and pK°_A are equal. The derivation of this equation is explained elsewhere (Lenhoff, 1968a).

The foregoing equation, which indicates that a plot of pK_A against pH will consist of a series of straight lines joined by short curved parts, holds true for the glutathione-hydra system (Fig. 3). The results follow almost exactly the predictions from the modified Dixon equations. The following interpretations were made (Lenhoff, 1969). (a) Ionizable groups at the receptor site participated in binding glutathione, because significant variations in pK_A occurred with change in pH. (b)

The concave downward inflections at pH 4·6, 4·8, 6·5, and 7·6 represented pK's of ionizable groups at the receptor site. These pK's probably do not represent ionizable groups of glutathione, which have pK's either below pH 4 (2·1 and 3·5) or above pH 8 (8·7 and 9·6) (Wieland, 1954). If the receptor site is protein, then the determined pK's may represent two β-carboxyls of peptide aspartic acid (or γ-carboxyls of peptide glutamic acid), an imidazole group, and a terminal α-amino group, respectively. (c) The horizontal lines indicate

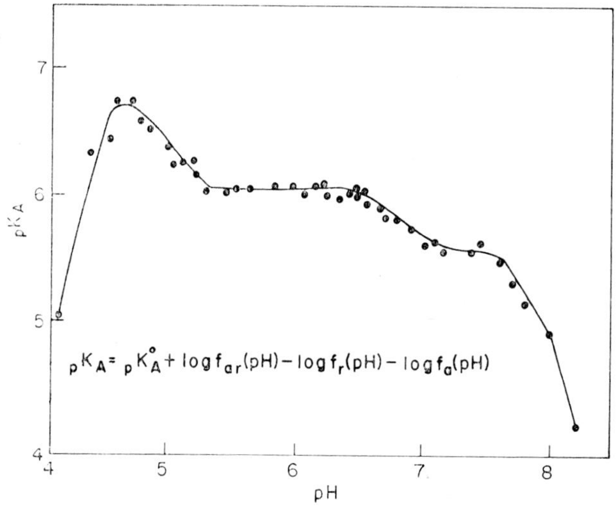

Fig. 3 Effect of pH on the dissociation constant K_A between glutathione and its receptor. (Reprinted from *American Zoologist*, with permission.)

pH values which do not effect the combination of glutathione with the receptor site. (d) the quenching of the charges (Dixon and Webb, 1958) at around pH 4 and pH 8 indicated that receptor-site groups having pK's of 4·6 and 7·6 may be associated with complementary charged groups on glutathione.

6. *Proposed Mechanism of Binding and Activation*

A proposed mechanism for binding of glutathione to the receptor site suggests that the charged groups at the receptor site bind complementary charged groups on glutathione. This proposal takes into account previous data which show that the receptor recognizes the tripeptide backbone of glutathione and that the free α-amino of the glutamyl moiety of glutathione is implicated in binding to the receptor. Thus, the positively charged α-amino of glutathione might neutralize a negatively charged carboxyl of the receptor, while the terminal carboxyl of the glycyl moiety of glutathione might bind to a positively charged group of the receptor's terminal α-amino. Similarly, the groups represented by pK's at pH 4·8 and 6·5 may be involved in the binding, or may be sufficiently close to the receptor site to be displaced somewhat during the binding process. These displacements are represented by the concave upward bends at pH 5·2 and 7·0 (Fig. 3).

The proposed binding mechanism points out the rigid specificity of the receptor for glutathione but does not tell us what happens after the combination occurs. Since, during activation, there was no detectable chemical alteration of glutathione and glutathione had to be constantly present at the receptor site, it was conjectured that glutathione operates by causing a reversible modification (possibly allosteric) of the tertiary structure of the receptor, which renders the receptor active (Lenhoff, 1961a; 1969).

IV. Evolutionary Considerations

A. Classification of Glutathione Response

The demarcation between behavior and physiology is not sharp. Although behavior is categorized by many scientists as an aspect of physiology, the distinction between these two fields cannot be put off as simply a matter of semantics. The mere categorization of a phenomenon often determines how the problem will be investigated, and by whom—that is, what experimental approach will be taken, and whether the investigator will be a psychologist or a biochemist, an ethologist or a physiologist. In the case of the lower aquatic invertebrates, the line between behavior and physiology becomes gossamer-thin because many lack well-defined nervous systems, endocrine systems, and structures usually associated with more highly organized forms.

In animals as primitive in organization as hydra it is difficult to classify the glutathione-activated behavioral response of feeding in the commonly used terms derived from extensive research done mostly with vertebrates and insects. Can we, for example, say that the actions of glutathione on hydra fall within accepted definitions for olfactory and gustatory phenomena? Are some of the actions of glutathione similar to those initiated by hormones or by hormone-like substances? Or must we revise or simplify our accepted definitions for chemical receptors to include the unique features of coelenterate feeding-activator systems?

Consider some of the characteristics and peculiarities of the glutathione-activated response, or for that matter, any response activated by a simple specific compound. In hydra the activation of feeding takes place in response to the size and shape of a single molecule; hence, if we were to call the glutathione response a gustatory phenomenon, it would be a unique one in which no coding or filtering mechanism is involved—one which, instead, would consist of an all-or-none activation of a specific type of receptor site. Since glutathione courses through an aqueous solution, it is inappropriate to use terms reserved strictly for olfaction. Furthermore, when we think of an organism detecting a compound present in the environment, we do not usually envision that organism as playing an active role beforehand in releasing that compound from an environmental source. Yet such is the case with hydra. The glutathione response is a secondary behavioral event that normally occurs only after another receptor-effector system has been activated. During feeding, hydra does not even become exposed to glutathione until a small aquatic organism accidentally contacts the hydra's tentacles and the nematocysts in the tentacles discharge, puncturing the now captured prey. Then, and only then, is glutathione released from the prey into the environment.

The action of glutathione might be considered to represent, in hydra, a coordinating mechanism akin to some endocrine systems of higher metazoans, because glutathione initiates the coordination of a complex series of contractions and relaxations and because, like hormones, it acts in low concentrations (10^{-9}M (Rushforth and Hofman, 1972) to 10^{-7}M (Lenhoff, 1965a). This view was taken by Loomis (1955) in his original article, where he emphasized that glutathione starts the machinery which coordinates the manipulation of the captured prey for ingestion.

The chemical control systems of hydra also seem to be as thoroughly integrated with the physiology of the organism as the hormonal systems of higher forms are. Discovery of the tyrosine enteroreceptor (Blanquet and Lenhoff, 1968) provides an example of a chemically mediated receptor-effector system that cannot exert its action unless another receptor, the glutathione one, is operating. Furthermore, Rushforth's demonstration (1964) that glutathione inhibits the hydra's contraction responses to either light or mechanical stimulation shows the influence of one chemically mediated receptor-effector system over two other receptor-effector systems.

Yet it is apparent that we cannot call the glutathione response hormonal, because: (1) the activator molecule comes from an organism of another species rather than from a gland within the organism itself; (2) the fluid environment that transmits the glutathione is external rather than internal; (3) the coordinated contractions are a visible response of the entire animal (hence termed behavioral) rather than a response of specific internal organs (hence termed physiological).

Were we looking for a name, glutathione, as it acts on hydra, might be called a "parahormone", a term "used to indicate conveniently those agents which act like hormones but do not entirely satisfy the accepted definition" (Gorbman and Bern, 1962). Some might even call it an "ectohormone". The danger of assigning a special label to a difficult-to-categorize phenomenon in a lower form of organism is that the name might overemphasize the unique aspects and hence obscure those features of chemical receptor mechanisms common to such systems in all metazoans. For example, despite the vast differences in complexity between hydra and mammals, the basic cellular events of receptor activation initiated by glutathione in hydra and by oxytocin in the uterus are probably similar. Both activators are relatively simple peptides that initiate coordinated responses in contractile tissue.

Thus we are left with the self-evident conclusion that, common to olfactory, gustatory, and hormonal (including ectohormonal) phenomena, is the following series of fundamental events: the combination of a cellular receptor with a specific molecule, leading to the initiation of a coordinated series of biological processes. Viewed in this light, the mechanism of activation is independent of such factors as whether the activating molecule issues from the prey or from an endocrine gland; whether the molecule traverses the external aqueous environment or the internal blood stream; whether the response is a behavioral contraction, an electrophysiological signal, or a sequence of developmental changes; or whether the phenomenon is classified as olfactory, gustatory, hormonal, or ectohormonal.

B. Evolution of Receptors for Specific Chemicals

The view that an organism evolves receptor sites in response to some ubiquitous molecules which are adapted to special tasks seems especially applicable to the feeding activators of coelenterates. So long as a molecule is widely present in prey organisms and has properties distinguishing it from closely related substances, it might serve as a feeding activator. One might therefore, expect some coelenterates to have evolved receptor sites for compounds other than glutathione emitted by the captured prey. Such is the case. As discussed in Section III, about 6 different amino acids have been shown to effect feeding in a wide variety of coelenterates.

After a receptor site for a specific compound had been acquired during evolution, further modification of the receptor site itself might have occurred. For example, Fulton (1963) has suggested that the evolution of a receptor site for glutathione into one for the α-imino acid proline may have proceeded by means of slight structural changes in the receptor site. He postulated this because one of the possible cyclized forms of glutathione in solution is close in structure to an α-imino acid. And, since proline is also present in the fluids released from prey organisms, the change in structure of the receptor site was not disadvantageous to *Cordylophora* but, under some circumstances, advantageous, and so persisted. Perhaps the zoanthid *Palythoa psammophilia*, which responds synergistically to a mixture of glutathione and proline (Reimer, 1971b), and the coral *Cyphastrea*, which responds to either glutathione or proline (Mariscal and Lenhoff, 1968) represent forms retaining "intermediate" forms of the receptors during the evolution of a purely proline receptor from a glutathione one.

We find it reasonable to suppose that receptor sites for amino acids or glutathione in hydra, and for transmitter substances and peptide hormones in higher organisms, may have evolved in a similar fashion. But the intriguing question is, "Did chemical receptors to environmental compounds, such as those found in lower invertebrates, give rise to the chemical receptors (olfactory, gustatory, hormonal, or transmitter) in higher forms?" For example, because dopamine and norepinephrine are formed from tyrosine, would it not be efficient for organisms to retain and utilize modifications of a primitive tyrosine receptor similar to the one found in hydra? Would not the same argument apply to the evolution of receptors for the neuro-transmitters glutamic acid and glycine? Such an argument could be taken back one more step into a consideration of chemical receptors for compounds stimulating pinocytosis in single cells. These receptors may truly be the evolutionary precursor of many receptors in higher forms.

Obviously, such evolutionary questions are virtually unanswerable. Nevertheless, perhaps there may be an argument in support of this evolutionary view, at least in the case of peptide receptor sites. In both lower and higher forms are receptor sites for specific peptides, which, when activated, lead to contraction responses. It would seem simpler for organisms, during evolution, to have modified existing receptor-effector systems to perform new tasks than to have developed a completely new receptor-effector system.

Such speculations, like most speculations about evolution, cannot be proved, but they may help to make us aware that unifying concepts, tacitly assumed in the case of enzymes and cell organelles, also may apply to the basic aspects of chemoreception. And, specifically, such speculations emphasize that the behavioral responses of lower invertebrates to a peptide or an amino acid and some hormonal and neurotransmitter responses in man may have many fundamental features in common. By focusing on the primary events of the combination of the activator with the receptor to initiate a series of coordinated activities, we may find new experimental organisms, new approaches, and new insights into universal, yet little understood, chemical control mechanisms.

V. Concluding Remarks

We have presented examples of how chemoreception plays important roles in the life history of many aquatic invertebrates. In particular we have indicated how the study of chemoreception involved in the feeding of coelenterates may provide a model to help us understand the basic mechanisms of action of one type of chemoreceptor. Also, by focusing on the basic events involved in all chemical receptor systems, we point out a possible path for the evolution of such receptors.

But what about future research in the study of chemoreception in aquatic invertebrates? We feel that because of the rapid development of sophisticated chemical separation and identification techniques during the past decade, the research will follow two lines. For one, we predict that an increasing number of compounds, mostly small molecules of known structure, will be shown to act as activators of chemoreceptors and that an increasing number of biological events, especially in the lower invertebrates, will be shown to be controlled by these compounds.

The excitement of the future, however, appears to us to lie in the actual identification of molecular receptors, their isolation, and the determination of their precise mechanism of action. Such techniques as affinity labeling and affinity chromatography, now being applied to the study of the acetyl choline receptor and the insulin receptor of higher forms, should prove extremely useful. Hopefully, just as the nerve of the squid axon provided unique experimental material for the major breakthrough in our understanding of nerve action, likewise, some lower invertebrate with a highly developed and experimentally accessible chemoreceptor system may provide suitable experimental material for some of the breakthroughs in our understanding of chemoreception.

References

Abe, N. (1938). *Palao trop. biol. Stn Studies* **3**, 469–521.
Adler, J. (1966). *Science, N.Y.* **153**, 708–716.

Adler, J. (1969). *Science, N.Y.* **166**, 1588–1597.
Araki, G. S. (1965). *Diss. Abstr.* **23**, 4306.
Bailey, D. F. and Laverack, M. S. (1963). *Nature, Lond.* **200**, 1122–1123.
Barber, S. B. and Hayes, W. F. (1963). *Proc. Int. Congr. Zool.* **16**, 76–78.
Beck, S. D. (1965). *A. Rev. Ent.* **10**, 207–232.
Beidler, L. M. (1954). *J. gen. Physiol.* **38**, 133–139.
Beutler, R. (1924). *Z. vergl. Physiol.* **1**, 1–32.
Blanquet, R. S. and Lenhoff, H. M. (1968). *Science, N.Y.* **159**, 633–634.
Brown, A. C. and Noble, R. G. (1960). *Nature, Lond.* **188**, 1045.
Carr, W. E. S. (1967a). *Biol. Bull. mar. biol. Lab. Woods Hole* **133**, 90–105.
Carr, W. E. S. (1967b). *Biol. Bull. mar. biol. Lab. Woods Hole* **133**, 106–127.
Carthy, J. D. (1958). "An Introduction to the Behaviour of Invertebrates". Allen and Unwin, London.
Case, J. (1962). *Comp. Biochem. Physiol.* **6**, 47–56.
Case, J. (1964). *Biol. Bull. mar. biol. Lab. Woods Hole* **127**, 428–446.
Case, J. and Gwilliam, G. F. (1963). *Proc. Int. Congr. Zool.* **16**, 75.
Case, J., Gwilliam, G. F. and Hanson, F. (1960). *Biol. Bull. mar. biol. Lab. Woods Hole* **119**, 308.
Castilla, J. C. (1972a). *Mar. Biol.* **12**, 222–228.
Castilla, J. C. (1972b). *Mar. Biol.* **15**, 236–245.
Castilla, J. C. and Crisp, D. J. (1970). *J. mar. biol. Ass. U.K.* **50**, 829–847.
Chapman-Andreson, C. (1962). *C. r. Trav. Lab. Carlsberg* **33**, 73–264.
Chernin, E. (1970). *J. Parasitol.* **56**, 287–296.
Cliffe, E. E. and Waley, S. G. (1958). *Nature, Lond.* **182**, 804–805.
Cohn, Z. A. and Parks, E. (1967a). *J. exp. Med.* **125**, 213–232.
Cohn, Z. A. and Parks, E. (1967b). *J. exp. Med.* **125**, 457–466.
Cooper, A. (1968). *C. r. Trav. Lab. Carlsberg* **36**, 385–403.
Coward, S. J. and Johannes, R. E. (1969). *Comp. Biochem. Physiol.* **29**, 475–478.
Crisp, D. J. (1967). *Biol. Bull. mar. biol. Lab. Woods Hole* **133**, 128–140.
Dixon, M. (1953). *Biochem. J.* **55**, 161–170.
Dixon, M. and Webb, E. C. (1958). "Enzymes". Longmans, London and New York.
du Vigneaud, V. and Miller, G. L. (1936). *J. biol. Chem.* **116**, 469–483.
Dragesco, J. (1962). *Bull. biol. Fr. Belg.* **96**, 123–167.
Ewer, R. F. (1947). *Proc. zool. Soc. Lond.* **117**, 365–376.
Faure–Fremiet, E. (1967). *In* "Chemical Zoology. I—The Protozoa" (M. Florkin and B. T. Scheer, eds.). Academic Press, London and New York.
Frings, H. and Frings, C. (1965). *Biol. Bull. mar. biol. Lab. Woods Hole* **128**, 211–217.
Fulton, C. (1963). *J. gen. Physiol.* **46**, 823–837.
Gorbman, A. and Bern, H. A. (1962). "A Text Book of Comparative Endocrinology". Wiley, New York.
Graham, R. A. and Mangum, C. P. (1969). *Va J. Sci.* **20**, 20.
Gurin, S. and Carr, W. E. (1971). *Science, N.Y.* **174**, 293–295.
Hashimoto, Y., Konosu, S., Fusetani, N. and Nose, T. (1968). *Bull. Jap. Soc. scient. Fish.* **34**, 78–83.

Hodgson, E. S. (1958). *Biol. Bull. mar. biol. Lab. Woods Hole* **115**, 114–125.
Jahan-Parwar, B. (1969). *Am. Zool.* **9**, 1097.
Jennings, J. B. (1960). *Biol. Bull. mar. biol. Lab. Woods Hole* **119**, 189–196.
Josefsson, Jan-Owe (1968). *Acta Physiol. Scand.* **73**, 481–490.
Kohn, A. J. (1956). *Proc. natn. Acad. Sci.* **42**, 168–171.
Laverack, M. S. (1963). *Comp. Biochem. Physiol.* **8**, 141–151.
Lenhoff, H. M. (1961a). *In* "The Biology of Hydra" (H. M. Lenhoff and W. F. Loomis, eds.). University of Miami Press, Coral Gables.
Lenhoff, H. M. (1961b). *J. gen. Physiol.* **45**, 331–344.
Lenhoff, H. M. (1965a). *Am. Zool.* **5**, 515–524.
Lenhoff, H. M. (1965b). *Nature, Lond.* **207**, 1003.
Lenhoff, H. M. (1966). *J. exp. Zool.* **163**, 151–155.
Lenhoff, H. M. (1968a). *In* "Chemical Zoology" (M. Florkin and B. T. Scheer, eds.), Vol. II. Academic Press, New York.
Lenhoff, H. M. (1968b). *Science, N.Y.* **161**, 434–442.
Lenhoff, H. M. (1969). *Comp. Biochem. Physiol.* **28**, 571–586.
Lenhoff, H. M. and Bovaird, J. (1959). *Science, N.Y.* **130**, 1474–1476.
Lenhoff, H. M. and Bovaird, J. (1960a). *Expl Cell Res.* **20**, 384–394.
Lenhoff, H. M. and Bovaird, J. (1960b). *Nature, Lond.* **187**, 671–673.
Lenhoff, H. M. and Bovaird, J. (1961). *Nature, Lond.* **189**, 486–487.
Lenhoff, H. M. and Brown, R. D. (1970). *Lab. Anim.* **4**, 139–154.
Lenhoff, H. M. and Schneiderman, H. A. (1959). *Biol. Bull. mar. biol. Lab. Woods Hole* **116**, 452–460.
Lenhoff, H. M. and Zwisler, J. (1963). *Science, N.Y.* **142**, 1666–1668.
Lenhoff, H. M., Muscatine, L. and Davis, L. V. (eds.) (1971). "Experimental Coelenterate Biology". University of Hawaii Press, Honolulu.
Levandowsky, M. and Hodgson, E. S. (1965). *Comp. Biochem. Physiol.* **16**, 159–161.
Lindstedt, K. J. (1971a). *Comp. Biochem. Physiol.* **39A**, 553–581.
Lindstedt, K. J. (1971b). *Science, N.Y.* **173**, 333–334.
Lindstedt, K. J., Muscatine, L. and Lenhoff, H. M. (1968). *Comp. Biochem. Physiol.* **26**, 567–572.
Lineweaver, H. and Burk, D. (1934). *J. Am. chem. Soc.* **56**, 658–666.
Loomis, W. F. (1955). *Ann. N.Y. Acad. Sci.* **62**, 209–228.
Loomis, W. F. and Lenhoff, H. M. (1956). *J. exp. Zool.* **132**, 555–574.
Mackie, A. M. (1973). *Mar. Biol.* **21**, 103–108.
Mackie, A. M. and Shelton, R. G. J. (1972). *Mar. Biol.* **14**, 217–221.
Mackie, G. O. and Boag, D. A. (1963). *Pubbl. Staz. zool. Napoli* **33**, 178–196.
Mangum, C. P. and Cox, C. D. (1966). *Am. Zool.* **6**, 546–547.
Mariscal, R. N. (1971). *In* "Experimental Coelenterate Biology" (H. M. Lenhoff, L. Muscatine and L. V. Davis, eds.). University of Hawaii Press, Honolulu.
Mariscal, R. N. and Lenhoff, H. M. (1968). *J. exp. Biol.* **49**, 689–699.
McLeese, D. W. (1970). *J. Fish. Res. Bd Can.* **27**, 1371–1378.
Mugard, H. (1949). *Annls Sci. Nat. Zool.* **10**, 171–268.
Nachmias, V. T. (1968). *Expl Cell Res.* **51**, 347–361.
Nagel, W. (1892). *Zool. Anz.* **15**, 334–338.

Pantin, C. F. A. and Pantin, A. M. P. (1943). *J. exp. Biol.* **20**, 6–13.
Pardy, R. L. and Lenhoff, H. M. (1968). *J. exp. Zool.* **168**, 197–202.
Passano, L. M. and McCullough, C. B. (1962). *Proc. natn. Acad. Sci. U.S.A.* **48**, 1376–1382.
Poilvert, A. (1959). *J. Protozool.* **6**, 118–130.
Reimer, A. A. (1971a). *Comp. Biochem. Physiol.* **40A**, 19–38.
Reimer, A. A. (1971b). *Comp. gen. Pharmac.* **2**, 383–396.
Rushforth, N. B. (1965). *Am. Zool.* **5**, 505–513.
Rushforth, N. B. (1967a). *J. gen. Physiol.* **50**, 2484.
Rushforth, N. B. (1967b). *Biol. Bull. mar. biol. Lab. Woods Hole* **133**, 449–450.
Rushforth, N. B. (1971). *Biol. Bull. mar. biol. Lab. Woods Hole* **140**, 255–273.
Rushforth, N. B. and Burke, D. S. (1971). *Biol. Bull. mar. biol. Lab. Woods Hole* **140**, 502–519.
Rushforth, N. B. and Hofman, F. (1972). *Biol. Bull. mar. biol. Lab. Woods Hole* **142**, 110–131.
Rushforth, N. B., Krohn, I. T. and Brown, L. K. (1964). *Science, N.Y.* **145**, 602–604.
Shelton, R. G. J. and Mackie, A. M. (1971). *J. exp. mar. Biol. Ecol.* **7**, 41–49.
Waley, S. G. (1958). *Biochem. J.* **68**, 189–192.
Wieland, T. (1954). *In* "Glutathione" (S. Colowick, ed.). Academic Press, New York.
Wieser, W. (1956). *Limnol. Oceanogr.* **1**, 274–285.
Wood, L. (1968). *Malacologia* **6**, 267–320.
Woodruff, L. L. and Spencer, H. (1922). *J. exp. Zool.* **35**, 189–205.
Zafiriou, O., Whittle, K. J. and Blumer, M. (1972). *Mar. Biol.* **13**, 137–145.

177–266

5. Factors Influencing the Settlement of Marine Invertebrate Larvae

D. J. Crisp

Unit of Marine Invertebrate Biology, (Natural Environment Research Council), Marine Science Laboratories, Menai Bridge, Anglesey, U.K.

I. Introduction

A planktonic stage is found in the life cycle of bottom living marine invertebrates belonging to all the major phyla of the animal kingdom except the Nematoda and Gastrotricha. During its short free-living stage, the minute larva is at risk to a multitude of pelagic carnivores and is in danger of being engulfed by filter feeders. Towards the end of this hazardous chapter in its life history, it must re-establish contact with the adult habitat. Not surprisingly, mortality is generally believed to be very great during this stage, although on account of the difficulty of identifying and tracing discrete populations drifting at large in the sea, it is difficult to measure instantaneous mortality rates quantitatively (Cushing, 1963).

The larvae of some animals are short-lived, settling within only a few hours of their release. Almost inevitably, such larvae are shed in a well developed condition with the apparatus necessary for settlement already functional. Nutrient reserves are needed to accomplish the re-organization of tissues at metamorphosis and to survive the rigours of post-larval growth when the adult feeding organs may still be in a rudimentary state. Hence such larvae are termed "lecithotrophic". The production of lecithotrophic larvae is not a feature of particular phyla, except perhaps the ascidians, but occurs sporadically in most phyla and is characteristic of certain genera, such as *Spirorbis* (Polychaeta), *Bugula* (Ectoprocta) and *Clava* (Coelenterata). Because of their ready availability for settling experiments, such larvae have been widely studied in the laboratory. The majority of larvae and marine animals, particularly in warmer waters, are obliged to spend a comparatively long period in the plankton. Such larvae have yet to complete their development from the egg or primitive embryo to the settling stage and have not usually been endowed with the energy resources necessary to do so. They must therefore feed on the surrounding plankton and are termed "planktotrophic". Their longer tenure of planktonic life exposes them to greater hazards and therefore the parent must produce them in larger numbers to ensure survival. Examples of phyletic groups that usually have planktonic larvae are the lamellibranchs, Crustacea and Echinodermata.

One can recognize a series starting with animals that produce large numbers of planktotrophic larvae, through those which reproduce with smaller numbers of lecithotrophic larvae to those which have eliminated the larval stage and its attendant risks altogether and reproduce viviparously. At present, we cannot generalize which of these reproductive modes imposes the greatest drain on the energy reserves of the parent, but there can be little doubt that in terms of fecundity—the numbers of eggs produced in adult life—the planktotrophic mode makes the greatest and the viviparous mode the least demand. The degree of dependence of the young on its parent or on the planktonic environment often differs widely in closely related animals. Cases have been reported where the mode of reproduction in a single species varies seasonally (*Spio filicornis*—Mesnil and Caullery, 1917), or regionally (*Brachyptomia rissoides* (Haul)—Rasmussen, 1951), while many genera contain species showing a whole spectrum of reproductive habits. For example, the boreal representatives of the genus *Littorina* comprise two northern species, *L. saxatilis*, which is viviparous, and *L. obtusata*, which is oviparous and whose young develop to the swimming stage in a gelatinous capsule; and two warm temperate species, *L. littorea* and *L. neritoides*, which produce planktonic eggs developing into planktotrophic larvae. The eggs of the subtropical oyster, *Crassostrea*, are shed directly into the sea where they develop into planktotrophic larvae; those of the temperate species *Ostrea edulis*, though first incubated in the parent to the swimming stage, still require some ten days to feed and develop before being able to settle, while those of the antipodean species *Ostrea lutrea* are released ready to settle.

The mode of reproduction in epibenthic and benthic forms seems to bear little relation to the habitat. However, as the above examples illustrate and as Thorson (1950) has amply demonstrated, the mode of reproduction broadly depends on the latitudinal range of the species. Few animals from arctic regions have planktonic larvae, although those that do are often abundant and widely distributed (e.g. *Balanus balanus*, *Saxicava arctica*). A considerable proportion of animals from temperate waters have either lecithotrophic or planktotrophic larvae with periods of development of the order of from one to three weeks. The larvae of animals living in tropical seas are predominantly planktotrophic, often with a very long period of pelagic development (see below). No doubt the basis of Thorson's rule lies in the restricted season of primary production at higher latitudes. It adds to the hazards of predation of the feeding larva the hazard of malnutrition if spawning does not coincide with the incidence of algal blooms. One arctic species at least, *Balanus balanoides*, L., has a highly developed mechanism to achieve synchrony (Crisp, 1956; Barnes, 1957; Crisp and Spencer, 1958).

In spite of its disadvantages, the pelagic larval stage has been independently evolved many times and remains the rule rather than the exception among marine invertebrates. It must therefore confer some compensating advantages but what these are can at present be only speculation.

Although the planktotrophic larva appears to benefit from the food present in the plankton, the species as a whole can hardly gain energy as a result of this mode of reproduction. The energy gained from planktonic food will be restricted to the growth increment from egg to settled larva of those individuals that eventually grow to become future reproducing adults. For a stable population, this would be only two per female in contrast to the many thousands that perish. The energy of the female in its lifelong reproductive effort is therefore sacrificed prodigally for the benefit mainly of planktonic predators. Measurements of reproductive outputs have not yet been made on a sufficient diversity of organisms to show whether less expenditure of energy is needed to produce the numbers of larvae to ensure survival with a planktotrophic than with any other form of reproduction. The supposedly high level of mortality in the plankton would suggest that it is not. Thorson (1950) suggested that the planktotrophic element in the life history was a device to resist benthic predation. Benthic organisms without a long pelagic stage, and particularly those with direct development, have comparatively stable populations which remain in the same place and can be relied upon as food by predators. Those that live a double life, relying on recruitment from the plankton at each generation, will sometimes occur in one locality, and sometimes in another and will show great variation in population size according to the strength of the year class. Animals with such opportunistic life cycles would not so readily be utilized by a predator for its staple diet. It may be relevant here to note that some groups of sedentary animals that permanently occupy large areas of the substratum and are capable of reproducing asexually —sponges, corals, anemones and ascidians—are usually unpalatable to predators.

An obvious function of the planktonic larval phase is to disseminate the species. The distance over which individuals can spread during a single generation would be determined by the term of larval existence and the strength and direction of the currents in which it drifts. For short lived lecithotrophic larvae, the range of dispersal can hardly exceed that of a tidal excursion or a day's wind drift, but for the many planktotrophic larvae which can drift for two or three weeks, distances of 20–100 km are possible, sufficient to transport them across many intra-continental seas. The larvae of many cool temperate species have this potential but they cannot survive a trans-ocean passage (Thorson, 1961). However, the larvae of certain tropical species belonging to a variety of phyla, and the larvae of a few warm temperate forms such as the phyllosoma of the spiny lobster, can endure many months in the plankton before undergoing metamorphosis, and can therefore be transported across the deep ocean. Scheltema has called them "teleplanic" larvae, from the Greek "teleplanos" meaning far-wandering.

Three possible benefits can result from larval dispersal; a wider genetic exchange, the potential to colonize distant geographical areas, and the ability to occupy an impermanent habitat.

It is generally accepted that a species containing many genotypes and with ample opportunity for genetic exchange has more stability and adaptability than one in which inbreeding is frequent. It is likely to have a more diverse habitat and a wider geographical range with a variety of interbreeding genetic combinations. Species with limited genetic exchange more readily form sub-species, each of which occupies a limited territory or slightly different habitat. Although there is evidence that the above correlation holds true in the marine environment, it is difficult to disentangle cause from effect. The majority of marine species with a dispersal phase also have obligatory sexual mechanisms of genetic recombination and usually occupy large domains. On the other hand, certain genera, such as *Spirorbis*, and certain species, such as the barnacle *Balanus amphitrite*, which have rather short-lived larvae and are potentially self-fertile, show great local diversity and variability. The relationship between larval dispersion, geographical range and genetic diversity of marine species needs further analysis before the genetic advantages of larval dissemination can be assessed.

The genetic advantages of larval dissemination may sometimes be achieved by an initial fusion of settled larvae. Sponges, notable for their ability to reconstitute after mechanical disruption, sometimes produce larvae which display the same ability by fusing after settlement (Delage, 1892). H. V. Wilson (1907) and Henriken (quoted in the same paper) were among the first to observe this phenomenon but further reports based on laboratory work have appeared from time to time (von Ankel and Eigenbrodt, 1950; Rasmont, 1970). Burton (1949) produced evidence that fusions occurred naturally, since the clustered pattern of young growths of several species gave the appearance of swarming and simultaneous settlement of the larvae. Fry (1971) described the aggregation of swimming larvae of *Ophlitospongia seriata* in the laboratory, and found that subsequent fusion reduced individual mortality. He pointed out not only the trophic advantage in coalescence (cf. Rasmont, 1970), but also the genetic advantage of the greater mixture of types present within a specimen of an hermaphrodite sponge. In fact, the end product was analogous to the results of gregarious settlement in non-colonial forms. A similar example of fusion of larvae has been described in corals by Edmondson (1946); the swimming planulae of *Dendrophyllia manni* can adhere and will then settle as a group. Aggregations of free swimming larvae are not infrequently observed in laboratory dishes, but the cause is probably purely mechanical and not necessarily an indication that swarming would occur in nature (Crisp, 1962).

The second advantage of larval dispersal is the ability to colonize areas remote from the main population of a species. The possible range is determined by the length of the larval phase and the strength of the currents carrying it. A good example of a teloplanic larva is *Phillipia krebsii* (Gastropoda, Archetectonicidae) which extends through the North Atlantic gyre by "island hopping" (Scheltema, 1971). The Semper's larvae of certain corals are also believed to undergo trans-atlantic passages, but since they are stenothermal, they must remain in tropical waters. It is thought that the equatorial surface currents carry them westwards and the equatorial undercurrent carries them eastwards.

The mode of larval development may be important even when species are transported artificially by attachment to the hulls of ships or in shellfish layings. Those with planktonic larvae are least capable of forming a permanent bridgehead, though if they do so their spread can be rapid (e.g. the barnacle *Elminius modestus*, Crisp 1958). Those without a planktonic dispersal stage (such as oyster drills) can more readily initiate a breeding stock, but their spread is relatively slow. Thus the rare phenomenon of species immigration is not necessarily favoured by the existence of a planktonic phase.

It is surprising that many organisms lacking a planktonic stage, living in such habitats as the interstitial spaces of sand and the deep seas, can nevertheless be found in widely separated geographical locations, suggesting that on the evolutionary time scale a planktonic larvae is by no means essential for dispersal. Colman (1950) emphasizes the same point in relation to predatory intertidal gastropods which lack free swimming larvae but which have a wide geographical range.

The advantage of a dispersal mechanism should not be judged solely in terms of the distance carried. The marine environment provides an extraordinary diversity of niches capable of supporting sessile or slow moving forms which feed on suspended or sedimented materials. A few of these organisms have only generalized requirements and occupy large areas of the bottom; the majority, however, have specialized requirements. Many deposit livers have an extremely patchy distribution, being restricted to sands or muds with a particular granulometric composition, organic content, and very often a specialized micro flora. Many species have a biotic relationship with others which determines what sites can be occupied. They may require to live in the immediate vicinity or on the surface of a particular host or prey organism. Good examples are provided by the barnacle *Pyrgoma anglicum*, epizoic on the cup-coral *Caryophyllia*, or the tube-worm *Spirorbis corallinae*, epiphytic on the alga *Corallina officianalis*. Some have even more specialized sites: the acrothoracican barnacle *Trypetesa lampas* lives in the columella of a gastropod shell in use by a hermit crab, the stalked barnacle *Conchoderma* favours attachment to the sessile barnacle *Coronula* which, in turn, lives half buried in the epidermis of the hump back whale. Such specialized habitats clearly do not persist indefinitely and a dispersal mechanism is essential to the continuation of the species. Even a physically determined and apparently stable habitat, such as a suitable location on a rocky shore, can become so overcrowded that it ceases to be effectively available to the succeeding generation. It is therefore highly advantageous for the species to possess a dispersal mechanism through which it can seek out a less heavily colonized area. The role of the planktonic phase in enabling a species to take advantage of such scattered and transient habitats is of supreme importance to animals whose powers of exploration virtually disappear at metamorphosis. Without it, such organisms would be restricted to very stable habitats and each enclave would suffer the disadvantages that inbreeding and over-specialization entail.

There is an exact parallel in the evolution of disseminules as an insurance against the disappearance of the habitat and their possible evolution as an insurance against predation, as put forward by Thorson. In the terminology of McArthur and Wilson (1967) such organisms may be referred to as r-species: they have high fecundity, they are opportunistic, and their populations are usually unstable in the sense that recruitment is potentially variable. In contrast, species which omit the dispersal mechanism are essentially K-species. They do not require such a high reproductive ratio, they will tend to form more stable and enduring populations, and their mortality will depend to a greater extent on density dependent factors than on mere chance.

If the planktonic larva is primarily an adaptation to enable a species with limited adult motility to exploit a scattered or transient niche and to avoid local overcrowding, it is clearly essential that the larva should be provided with the mechanisms it needs to find and recognize the appropriate niche. Reliance on chance encounter at the moment the larva is about to metamorphose, as was once supposed (Colman, 1933), would be disastrously uneconomic for the majority of organisms since their habitats form such a small fraction of the available area.

The typical larvae of all phyla are provided, in addition to organs of swimming and attachment, with sense organs which enable them to prospect for and identify their habitat. Not surprisingly, larval sense organs reach a peak of development at the time of settlement and metamorphosis, and are frequently discarded or degenerate thereafter. Table 1 lists the settling larvae found in each major group indicating the sense organs known to be present and their subsequent fate. Not uncommonly, larvae are known to react to light, to gravity, to a specific chemical substance or to change of pressure, although the sense organ responsible has not yet been discovered. If these latent sense organs are included, it can be seen from the table that the great majority of larvae can detect the direction of light and gravity, and may also be capable of responding to light intensity, pressure, water currents and chemical substances. The best studied larvae have the largest repertoire, so no doubt further investigations will increase the known capacities of the others.

II. Larval Navigation in the Pelagic Phase

Since the planktonic larva has little ability to swim, at least at a rate significant in comparison with the rate at which it is carried by tidal

Table 1. Sense organs of marine invertebrate larvae at the settling stage

Phyletic group and footnote reference number	Sensitive to: Light direction	Light intensity	Gravity	Water flow	Surface contact	Pressure	Chemical substances	Attachment organ	Notes
Sponge larvae 4, 5, 6	(+)		(+)	(+)				Anterior, sparsely ciliated, pole	
Coral planulae 1	(+)		(+)	(+)	(+)			Nemato-cysts and glands at aboral end	
Hydroid planulae 2, 3	(+)			(+)	(+)		(+)		
Polychaete larvae 7, 8, 9, 10	+				(+)		(+)	Cement in serpulids and *Sabellaria*	*Hydroides* not sensitive to light until *both* eyes formed
Bivalve pedi-veligers 11, 12, 13, 14	+	+	(+)		+	(+)	(+)	Byssus apparatus	Eye normally lost in adult
Gastropod post-larvae 15, 16, 17	+		+		+		(+)	Plantar surface of foot	
Polyzoan larvae 19	+	+	(+)	(+)	(+)		(+)	Eversible sac	Eyes lost

Barnacle cyprids, 20	+	+		(+)	+	(+)	(+)	Antennulary attachment disc and cement apparatus	Paired compound eyes lost
Echinodermata								Fixation disc, permanent in crinoids	No modern work on settling stages
Ascidians 21, 22, 23	+	+	+					Adhesive papillae	Eyes and statocyst lost

+, Sense organ and response observed in the larva of at least one species quoted; (+), Response observed but sense organ not known.

Reference: 1, *Archelia horrescens* (Kawagati, 1941); 2, *Hydractinia echinata* (Teitelbaum, 1966); 3, *Clava squamata* (Williams, 1965); 4, *Microciona prolifera* (Warburton, 1966); 5, *Mycale macilenta*, *Haliclona* sp. (Bergquist and Sinclair, 1968); 6, *Ophlitaspongia seriata* (Fry, 1971); 7, *Spirorbis* sp. (Knight-Jones, 1951; Wisely, 1960; De Silva, 1962; Williams, 1964; Gee, 1965); 8, *Hydroides norvegica* (Wisely, 1958b); 9, *Ophelia bicornis* (Wilson, 1952, 1955); 10, *Sabellaria* sp. (Wilson, 1968, 1970a, b); 11, *Mytilus edulis* (Bayne, 1963, 1964a); 12, *Ostrea edulis* (Cole and Knight-Jones, 1939, 1949; Bayne, 1969); 13, *Crassostrea virginica* (Nelson, 1928; Prytherch, 1934; Crisp, 1967); 14, *Teredo* sp. (Isham *et al.*, 1951; Isham and Tierney, 1953; 15, *Nassarius obsoletus* (Scheltema, 1961); 16, *Adalaria proxima* (Thompson, 1958); 17, General review (Fretter and Graham, 1962); 18, *Bugula* sp. (Lynch, 1947, 1949b; Grave, B. H. 1930); 19, Various Polyzoa (Ryland, 1959, 1960, 1962a; Crisp and Williams, 1960; Wisely, 1958a); 20, *Balanus balanoides* (Barnes *et al.*, 1951; Crisp, 1953; Crisp and Barnes, 1954; Knight-Jones, 1953b; Knight-Jones and Qasim, 1966; Crisp and Meadows, 1962; Nott and Foster, 1969); 21, *Amaroucium* sp. (Mast, 1921; Grave, 1921); 22, *Styela partita* (Grave, 1944); 23, General review (Berrill, 1950).

streams and ocean currents, it might be thought to be quite incapable of controlling its movements and therefore to be at the mercy of the body of moving water in which it drifts. However, the capacity which some larvae possess to regulate the depth at which they float by adjusting the length of periods of upwards swimming and of passive sinking can provide them with some influence over their fate. Insofar as these mechanisms are of adaptive value, they may be described by the term "larval navigation".

The simplest and the most universal of such mechanisms is the response to light and/or gravity at the first swimming stage. Almost invariably, the newly emergent larvae swim vigorously upwards and towards the light (Thorson, 1964). By this means, they are quickly removed from the vicinity of benthic filter feeders which might engulf them and are brought as speedily as possible into the surface waters. For larvae with low visual thresholds, it would be necessary to phase the release to coincide with daylight and not infrequently natural release is known to occur at dawn. It is presumably triggered by increasing light intensity. In the laboratory, release can be delayed by keeping the developing young in the dark and stimulating them by exposure to light. Lecithotrophic larvae of many invertebrates may conveniently be obtained in quantities for experiment by keeping the parent colonies in covered dishes for a few days while the embryos develop to the hatching stage, and then exposing them to light. Swarms of larvae are often produced in this way and collect on the illuminated side of the dish. A positive response to light is characteristic of early planktonic larvae but not necessarily of larvae in general. The larvae of *Scoloplos armiger*, for example, which hatch directly into a benthic form within the egg capsule, crawl downwards through the neck of the capsule and into the deposit where they immediately develop (Thorson, 1946).

The planktonic larva does not necessarily continue to respond positively to light; its later behaviour is related to the habitat which it will ultimately occupy. Thorson (1964) correlated the photic behaviour at or near the time of settlement with the habitat, whether in the intertidal zone or in the shallow sublittoral. Larvae of invertebrates occupying the former were variable in behaviour; some remained photopositive, some became indifferent to light while some became photonegative. Although a photopositive larva might have the best chance of ensuring contact with an object in the intertidal zone, a short-lived larva which did not respond to light or indeed a larva with a late developing photonegative response, would not necessarily be disbarred from settling in a dark part of the intertidal zone. There are in fact four species commonly found on the fronds of *Fucus serratus*, sometimes on the very same frond, which display all possible responses to light at settlement (see later). Larvae of species living entirely in the sublittoral, however, are virtually unanimous in reversing the sign of the response to light when about to settle. Clearly, any larva which remained photopositive and maintained a position near the surface of the sea would be rendered incapable of occupying a habitat below the tide marks.

Ryland studied the development of the photonegative response, so characteristic of most polyzoa, using larvae of *Celleporella hyalina* and *Cryptosula pallassiana*. The beginning of the photonegative phase was accompanied by a greater tendency to explore the surface of algal fronds on which *Celleporella* eventually settled, and since larvae settled on the part of the frond most distant from the light, the photonegative phase must be a normal prelude to settlement. It can therefore be considered as a feature of adaptive value to sublittoral or shade loving species. Contrary to the implication of Lynch (1947), who studied the larva of *Bugula neritina*, the change in response to light owes nothing to the amount of illumination which the larva has received. It is quite different from the reversible adaptation to strong light shown by many larvae in the natant phase. The change in response of *Celleporella* is rather a developmental change which takes place during the first three or four hours of free swimming life (at 20°C) whether or not the larva is illuminated. Ryland further showed that in *Cryptosula pallassiana*, the rate of spontaneous reversal of the photic response increased with temperature and if his results are plotted in accordance with the Arrhenius equation they show a Q_{10} of about 4·5—a value quite similar to that shown by other developmental changes in poikilotherms.

More interesting examples of larval navigation have been demonstrated in benthic species living in estuaries. The residual downstream drift caused by freshwater flow poses a problem for estuarine invertebrates with planktonic larvae quite different from that facing anadromous fish: how can the larva, with no effective propulsion, swim against the current and return to colonize the upper reaches of the estuary where it was spawned?

Bousfield (1955) addressed himself to the study of the transport of larvae of the estuarine barnacle *Balanus improvisus* in the Miramichi estuary in Canada. He showed that although the early stages were photopositive and were carried downstream, the later naupliar stages remained close to the bottom and were transported upstream by the deeper saline water which contributes to the vertical circulation of a typical estuary (Pritchard, 1953).

No doubt a mechanism based on this principle could account for the persistence in estuaries of the majority of species with fairly long lived planktonic larvae, but further refinements are possible. As early as 1912, J. Nelson proposed a somewhat different mechanism to account for the upstream transport of bivalve larvae such as those of the American oyster *Crassostrea virginica.* He supposed that, stimulated by a combination of increasing salinity and current flow, the oyster larvae rose from the bottom and were carried in the water mass during the flood tide, but dropped to the bottom and lay still during the ebb. This hypothesis was supported by the observations of T. C. Nelson (1930, 1954) and those of Haskin (1964) who both claimed to demonstrate an increase in swimming activity correlated with an increase in salinity and current. Criticisms of the hypothesis that a pattern of swimming activity synchronous with tidal currents carries oyster larvae upstream have been made on the grounds that the mechanism may not be necessary. Korringa (1952) considered that an upstream drift by passive transport near the bottom was a sufficient explanation, and Manning and Whaley (1954) have also proposed more simple mechanisms which are reinforced by Carriker's (1951) and Kunkle's (1957) data. As with *Balanus improvisus* larvae, the younger *Crassostrea* veligers were distributed more uniformly throughout the water column whereas the older larvae remained nearer to the bottom. Recently, Wood and Hargis (1971) have revived Nelson's thesis, by showing in a large-scale field study that coal particles, which they claimed had densities and size ranges similar to those of oyster larvae, were distributed in time and space in a quite different manner. Whereas the maximum number of coal particles were brought into suspension four times a day corresponding, as expected, with the maximum tidal flow, the oyster larvae were present at maximum concentration only twice a day. The time at which the maximum numbers of oyster larvae appeared was not at the time of maximum flow in the flood tide, but later on, towards the time of high slack water, in agreement with earlier findings by Prytherch (1929). This evidence therefore supports the view that larval behaviour contributes to the efficiency of the upstream transport mechanism, but more experiments on larval swimming behaviour under simulated natural conditions are needed to clinch the matter.

III. Responses by Pelagic Larvae to Environmental Variables

Surprisingly little is known of the behaviour of free swimming larval stages. Their small size and, for the most part, their immediate response to light, makes them difficult to observe and experiment with in the large volumes of water needed to reproduce conditions in the open sea. *A priori* there are only three physical clues in the environment which have vector properties—light direction, gravity and water flow—and of these only the first two can be detected by the larva when it is out of range of points of reference. Light flux and gravity are both usually directed vertically downwards from the surface. All other environmental variables, namely light intensity, pressure, temperature and the concentration of dissolved substances (including salinity) are scalar quantities which can give no information to enable the larva to orientate unless their variation in space can be detected. Thus pressure can be used to determine the vertical only if it can be measured at several points and the direction of the pressure gradient established. The same is true of locating a chemical source by sensing its odour; the pattern of intensity variation must be detected in order to establish the direction of the chemical gradient leading to the source. Because the larvae are extremely small, and because sense organs are sensitive to logarithmic rather than linear differences, the chances are quite remote that a larva could register variations in the level of response between one sense organ and another on it surface. Nevertheless, directed responses induced by variations or thresholds in scalar quantities have been evolved. The expedient is to link the animal's response to a scalar quantity to its reaction to one of the vectors, light or gravity. Thus, a change in hydrostatic pressure or the detectable presence of a chemical entity in the environment can be used as the trigger bringing about a response to light. The crawling planulae of *Clava squamata* provide a good example of a directed response to a chemical stimulus. Williams (1965) observed that this larva was normally photopositive, but when in contact with the fronds of the brown alga *Ascophyllum nodosum*, the sign of its response to light was reversed, causing it to move into the darker recesses of the plant and to settle in the axils of the branches.

Responses to relatively small changes in hydrostatic pressure, of the order of 100 millibars or less, are widespread among marine invertebrate larvae, despite the absence as far as is known either of pressure sensing organs or of pressure sensitive structures such as gas vesicles. The response of a larva to increase in pressure is generally to swim upwards and the response to decrease in pressure to swim downwards or to sink passively under its own weight. Hence it seems likely that the adaptive significance of these responses is to regulate the level at which the larva maintains itself in the water column. Knight-Jones and Morgan (1966) have reviewed the responses of various organisms to change in pressure and have shown that some barosensitive larvae orientate to gravity and others to light. Those using gravity normally have well developed statocysts and swim upwards in response to increases in pressure. When the pressure falls, they may swin actively downwards—for example, the megalopa larvae of *Carcinus* or the juveniles of *Loligo forbesei*—or they may sink passively, as do the ephyra larvae of *Aurelia aurita* (Rice, 1964). The veliger larvae of *Mytilus edulis* swim upwards to the surface with increasing vigour as the pressure on them is increased; with decrease in pressure the velar cilia beat more lethargically and the larva tends to sink (Bayne, 1963). Larvae which orientate to light and normally show positive phototaxis may swim away from the light when the pressure is decreased, e.g. *Balanus balanoides* nauplii (Rice, 1964; Knight-Jones and Morgan, 1966) and the furcilaria larva of *Meganyctiphanes norvegica* (Rice, 1964), while according to Knight-Jones and Qasim (1966), the nauplii of *Elminius modestus* in the photonegative condition can be made photopositive after a positive pressure pulse but do not respond to negative pulses or to a fall in pressure. Other Crustacea, such as the cyprids of *Balanus balanoides* (Knight-Jones and Qasim, 1955) and the zoea of various decapods (Rice, 1964) sink passively with a fall in pressure or a negative pulse. Not all larvae can be made to respond to pressure; for instance the phyllosoma of *Palinurus elephas* (Rice, 1964), the tadpole larva of *Diplosoma listerianum* (Crisp and Ghobashy, 1971) and the larvae of the cephalopods *Sepiola atlantica* (Rice, 1964) and *Alloteuthis subulata* (Knight-Jones and Morgan, 1966) appear to be indifferent to pressure changes.

Despite the prevalence among invertebrate larvae of responses to small pressure changes, the sense organ and indeed the mechanism involved is at present unknown. None of the proposed mechanisms detailed in Knight-Jones and Morgan's review have much supporting evidence nor are they without serious objection. They mention the possible presence of invisible gas organs (Hardy and Bainbridge, 1951), pressure sensitive potentials in semi-conducting material (Digby, 1961a, b), pressure sensitive enzyme production, and a sensitivity to differential compression through skeletal proprioceptors or other sense organs. Most other barosensitive physiological phenomena occurring in liquid or solid phases involve pressure changes of at least an order of magnitude greater than those detectable by invertebrate larvae.

Light intensity, as distinct from light flux, resembles pressure in being a scalar quantity. Casual observations can demonstrate responses to light flux, but much more careful experimentation is needed to reveal the response to light intensity. In order to study the effect of spatial variations in intensity, it is necessary to provide a gradient of light intensity at right angles to the direction of the incident light (see Ryland, 1960). Even less attention has been given to experimental work to determine the effects of temporal changes of light intensity on swimming larvae, though it is known from non-quantitative observations that some larvae, such as the ascidian tadpole and the barnacle cyprid, are influenced by suddenly increasing or decreasing the level of illumination. Sudden increases in light intensity cause the larvae to sink or to swim away from the light, sudden decreases cause them to swim upwards or towards the general direction of the light source. These responses could explain the settlement often observed on shaded areas surrounded by a bright background, particularly of larvae which remain photopositive up to the time of settlement (see below, p. 222). Settlement may sometimes be advanced as a result of changes in light intensity; it is known, for example, that fluctuating light reduces the length of the free swimming stage in ascidian tadpoles (Grave, 1935; Crisp and Ghobashy, 1971).

Although there is little difficulty in understanding, at least superficially, how the eyes of most larvae function, few studies have been made of the relationship between visual input and locomotory movements. An exception is the classical study by Mast (1921) of the phototactic responses of the ascidian tadpole. The larva of *Amaroucium* was large enough for him to impale under a microscope and observe while artificially illuminated from either side. He noticed that the single eye had a rather narrow angle of acceptance of light and that the direction of the swimming movements executed by the tail was determined by the change in the light intensity received. If the light was increasing, it would turn in one direction, if the light was decreasing it would turn in the other. Thus when the free swimming larva was laterally illuminated it followed a spiral course, changing the direction of tail movement at every 180° rotation and thereby turning consistently either towards or away from the light according to its photopositive or photonegative state. Mast's analysis emphasizes the importance of spiral swimming, which is such a common feature not only of invertebrate larvae but also of many other small organisms. One may wonder whether the need to scan continuously the visual field and possibly also the gravitational field may lie at the root of such behaviour. It is arguable that for a small organism it might be easier to appreciate and interpret the resulting fluctuation in stimuli reaching a single sense organ during spiral swimming than to register the balance of stimuli received from two or more sense organs during steady movement.

The physical and chemical constitution of the sea water surrounding the larva may also influence its behaviour. As has been mentioned, rising salinity is believed by some authors to stimulate swimming in oyster larvae. Lowered salinity near the water surface may limit the migration of larvae and other planktonic organisms into the upper part of the water column if the laboratory observations of reduced upwards swimming (Lance, 1962) or the reversal of positive phototaxis (Ranade, 1957) at low salinities can be extended to a natural habitat with salinity stratification.

Temperature gradients in the water column may also determine the depth at which the larva maintains itself, though this conclusion similarly rests on the extrapolation of laboratory observations to the sea. Following Loeb's (1893) observations on *Polygordius* and marine copepods, Kanda (1919) found that a sufficient rise in temperature reversed the photopositive swimming of the larva of *Arenicola cristata* and Ewald (1912) found that the positive response of nauplius larvae can be similarly reversed by a rise in temperature. In contrast, he found that lowered oxygen levels and lowered pH promoted the photopositive response. Grave (1935) observed similar changes in ascidian tadpoles. In common with those of many other ascidian larvae, those of *Diplosoma listerianum* were shown by Crisp and Ghobashy (1971) to alternate in behaviour between periods of upwards swimming and periods of sinking. The periods of sinking were shown to be progressively increased under strong illumination or if the temperature exceeded 16°C. The authors suggested that this behaviour would exclude the larvae from highly illuminated surface waters and cause them to descend below the thermocline in warm weather. The free swimming veliconcha and eyed veliger larvae of *Mytilus edulis* show similar changes with temperature; temperatures above 15°C reduce both the geonegative and the photopositive response which would normally take the larvae higher in the water column (Bayne, 1964a). The response to light in the planktonic larvae of barnacles can be modified also by their nutritional state. Recently fed larvae tend more often to be photonegative and starved larvae become increasingly photopositive (Singarajah *et al.*, 1967), a relationship found in other planktonic organisms (Lucas, 1936) and readily interpreted as adaptive. Indeed, all the above responses can be interpreted in this way, since they cause the larvae to move in or out of the euphotic zone when conditions become extreme.

There appear to be at least two examples of a directed response of pelagic larvae to a specific chemical substance dissolved in the water. Scheltema (1961) demonstrated that the larvae of the mud snail *Nassarius obsoletus* (Say), an abundant scavenger of estuarine intertidal sand and mud flats, could delay metamorphosis for some two weeks if a suitable substratum was not encountered. If larvae were immersed in sea water which had stood over a layer of the natural substratum for several days, they were stimulated to settle. The active principle was evidently soluble as it was retained after the water had been passed through a millipore filter of pore size 0·8μm. Scheltema believed that the swimming-crawling stage was driven up estuaries by residual currents near the bottom and would be induced to terminate its pelagic existence and metamorphose on the bottom deposits, having sensed the presence of the soluble factor in the water. The second example of a response of a pelagic larva to a dissolved chemical substance is drawn from the much earlier work of Harington (1921) on the larvae of the shipworm *Teredo norvegica*. In experiments on a laboratory scale, he found that these larvae were chemically attracted to extracts of wood, and he simulated the effects with a 1% solution of malic acid. However, he did not think that attraction of larvae up a chemical gradient was likely in the vast arena of the sea, but suggested that it would be advantageous to the larva to recognize the presence of wood when it came in contact with it. Wilson (1952) reported some preliminary observations on *Teredo* which showed that this was indeed the case. If kept in glass vessels *Teredo* larvae lived for a week but failed to metamorphose and died. But if placed in contact with a suitable substratum of wood or paper, they settled, metamorphosed and bored into it. Some molluscan larvae can therefore respond by means of their organs of chemical sense to the presence of substances in solution and can possibly show chemotaxis to the concentration gradient at close range.

IV. Delayed Metamorphosis

In the last section, two examples were given of substances which caused the larva to metamorphose if it encountered them and made contact with the appropriate substratum for settlement. Obviously, in experiments in which metamorphosis is induced the larva must be physiologically ready to metamorphose with the morphogenetic processes held in abeyance until the larva encounters the stimulus or series of stimuli associated with the habitat. At one time, it was generally thought (e.g. Nelson, 1928) that since larvae pass through their developmental stages according to a temperature-dependent timetable, metamorphosis might also be subject to an endogenously controlled schedule and therefore take place as soon as larval development was completed. In fact however, most larvae can remain at the pre-settlement stage for a more or less indefinite period. The possibility of delaying metamorphosis at the last larval stage impressed such biologists as Matthews (1917), Wilson (1932) and Cole (1938) who realized that it gave the larva a greatly extended period of time in which it might find a suitable habitat. This idea led to the rejection of the view that marine larvae must settle at random; and as mechanisms of site recognition became understood, it was accepted that pelagic larvae exercise considerable control over their place of settlement.

Despite the profound structural reorganization that is usually involved, once the larva has found a congenial habitat and metamorphosis has been initiated, it is usually completed in a remarkably short time. For example, barnacle cyprids require in general less than 24 hours and some, such as *Elminius modestus*, as little as 8 hours to become small barnacles. It is, in fact, those organisms which undergo the most far-reaching changes that metamorphose the most quickly. The reason is not difficult to appreciate. Where there are marked differences between the larva and the adult in the mode of life, behaviour or nutrition, corresponding differences are likely to be reflected in their morphology, and metamorphosis will need to be more radical. While the larva is undergoing these great changes in its structure, it is unlikely to be able to sustain or defend itself and it is therefore essential to minimize the time which it spends in an intermediate form. It is well established that many larvae cease to feed during metamorphosis, thereby imposing a nutritional stress which must make demands on stored food reserves. Larvae which have a free swimming existence of only a day or two will be able to carry sufficient reserves from the egg to tide them over metamorphosis, but larvae with longer free living stages will be dependent on conditions in the plankton for obtaining sufficient food. It is widely believed that when the plankton bloom is poor or delayed, the crisis of metamorphosis may take a considerable toll of wild larvae, leading to failure in spatfall—a hazard in the management of commercial oyster and other bivalve industries. Hence a great improvement in production may become possible by artificially hatching larvae under good conditions of growth. The risk of insufficient food for planktotrophic larvae is particularly high in the arctic owing to the short period of plankton production. One arctic species at least, *Balanus balanoides*, has evolved a delicate mechanism, mediated by a pheromone-like substance, to ensure that the nauplii are not released until planktonic food appears in the water (Crisp, 1956; Crisp and Spencer, 1958).

The cirripedes, which are perhaps entitled to the highest place in the evolution of sedentary animals, illustrate particularly well the importance of food reserves and delayed development. Evidently, the advantages of dispersal are so great that a distinctive larval stage intervening between the feeding nauplius and the metamorphosed adult has been evolved. This is the cyprid, which has no trophic function but is beautifully adapted with specialized sense organs and attachment mechanisms to prospect for a suitable habitat (Walley, 1969; Nott and Foster, 1969; Nott, 1969). Darwin appropriately referred to the cyprid as a "pupa", but, unlike the corresponding lepidopteran stage, the cyprid is highly active, swimming for a term of up to several weeks. Its gut is non-functional, the nauplius stages having accumulated large stores of lipid globules, which can be seen occupying much of the anterior part of the cyprid.

A number of workers have described abnormally rapid metamorphosis in response to sub-lethal doses of toxic substances and other abnormal conditions. For example, Huxley (1928) accelerated metamorphosis in *Echinus plutei* with very dilute mercuric chloride, Prytherch (1934) induced metamorphosis in *Crassostrea virginica* and Grave (1944) in the ascidian *Styela*, with copper salts. Ascidian metamorphosis appears to be extremely labile, other accelerators being solutions of ethanol (Ghobashy, 1967), vital dyes (Bradway, 1936; Bertholf and Mast, 1944; Zhinkin, 1938), abnormal ionic media (Bradway, 1936) and overcrowding (Grave, 1944). Metamorphosis in the bryozoan *Bugula neritina* can be accelerated by hypertonic solutions, copper salts, oxidizing agents, vital dyes and X-radiation of the sea water (Lynch, 1949a, 1952). According to Grave, the ascidian tadpole is a dual organism with adult structures present in a rudimentary form. He believed that the development of the adult was inhibited by the activity of the larva. Hence he postulated that poisoning or crowding or other deleterious conditions, by reducing the metabolic activity, might release the developmental processes of the adult from dormancy. But there is no independent evidence to show that larval structures are more sensitive than developing adult structures to such poisons. Nor is there any direct evidence that metamorphosis induced by crowding is of adaptive value in causing gregarious settlement. Lynch (1952, 1956, 1958, 1960) in a series of papers on the effects of various treatments on polyzoan and ascidian metamorphosis, thought that the inactivation of an enzyme system such as succinic dehydrogenase might initiate metamorphosis. However, urethane, which should also inhibit succinic dehydrogenase, not only failed to induce metamorphosis but, in common with other anaesthetic agents, was itself inhibitory.

Perhaps there is no common factor in the action of various agents which induce or suppress metamorphosis, but it seems clear that they must act directly on, and modify the metabolic processes of, the larva. In contrast, when the larva encounters the hierarchy of stimuli indicative of the adult habitat, neural and possibly hormonal communication sets in train the normal development of the adult structures.

An abnormal situation of a different kind arises when larvae are denied access to a suitable habitat for an excessive period of time. As described below, the stimulus required for metamorphosis gradually diminishes with age, and the pre-settlement behaviour pattern becomes shorter or is eliminated, so that the ability to discriminate between habitats may become lost. It appears that, beyond a certain time characteristic of the particular larvae, the developmental processes can no longer be dammed back and the larva proceeds to metamorphose under quite inappropriate conditions. Such metamorphoses may also be abnormal: Wilson (1932) and Knight-Jones (1953a) described such abnormalities in larvae of *Owenia fusiformis* and *Spirorbis borealis* respectively. In extreme cases the larvae may lose its capacity to metamorphose altogether, as Knight-Jones (1953b) observed in barnacle larvae, Williams (1964) in *Spirorbis* larvae, and Wilson (1968) in *Sabellaria* larvae.

V. Behaviour at Settlement

The settlement behaviour of larvae belonging to entirely unrelated phyla presents surprising uniformity. The order of events is predictable and unremarkable: temporary attachment on making contact with the substratum followed by exploration during which the larva prospects the surface and finally, if the substratum proves suitable and the larvae does not swim off, metamorphosis and fixation. Metamorphosis marks the end of the pelagic stage and in the case of a sedentary species it is associated with permanent fixation by means of an adhesive often in a chosen orientation which improves its chances of survival (Barnes *et al.*, 1951; Crisp, 1953; Crisp and Barnes, 1954; Forbes *et al.*, 1971). In burrowing and crawling forms, the larval feeding and locomotory systems are usually lost but sometimes, as in bivalve molluscs, a swimming-crawling stage, termed the pediveliger (Carriker, 1956), intervenes. This intermediate stage performs the same searching role as the cyprid in barnacles, but is less obviously specialized and generally of shorter duration.

It is the sequential pattern of exploratory movements that demonstrates convergence to a degree. Though the different elements in behaviour merge together, there are broadly three phases which can be identified. The first may be called "broad exploration". On alighting, the larva takes a more or less straight path, turning rather infrequently. If it remains long enough on the surface, its behaviour changes to the second pattern which may be termed "close exploration". The path is made up of short steps with frequent changes of direction. Finally, as metamorphosis nears, the larva enters a third phase, scarcely translating at all but rotating and moving to and fro within its own length. This last phase may be called "inspection", since the larva appears to be minutely investigating, and possibly even preparing, the surface of the immediate area to which it will cement itself. Tracks taken by barnacle, tubeworm, and oyster larvae are illustrated in Fig. 1; the similarity is obvious and the three phases can be easily recognized. In the barnacle cyprid, the changes in its behaviour do not depend on any change in the means of locomotion. It literally "walks" by placing the adhesive antennules alternately one in front of the other. As the settlement pattern matures, only the degree of turning and twisting increases, so that

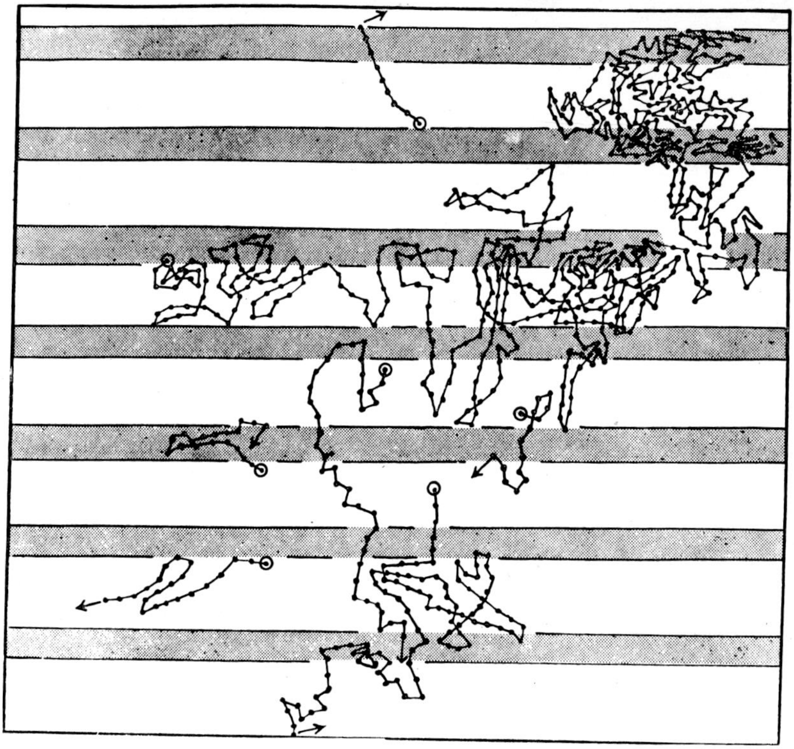

Fig. 1(a)

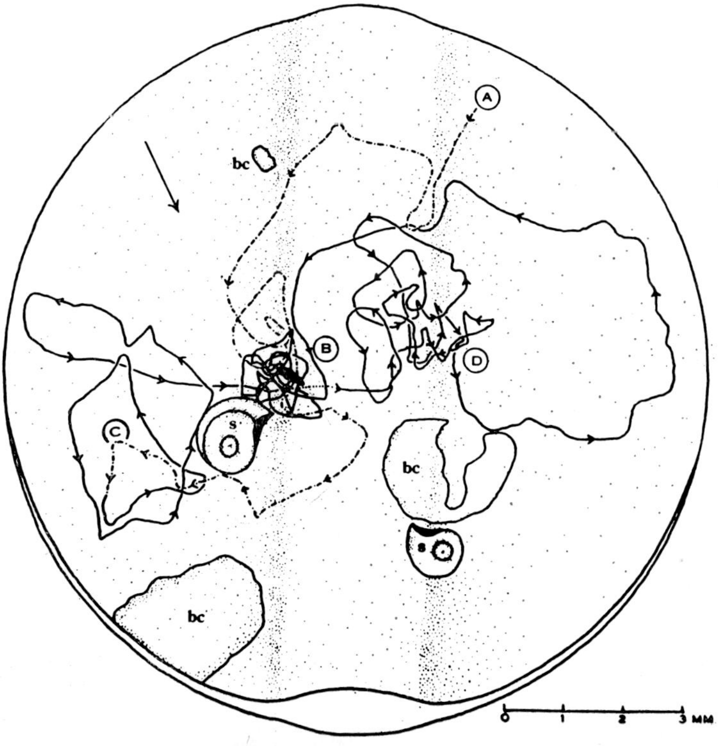

Fig. 1(b)

Fig. 1(c)

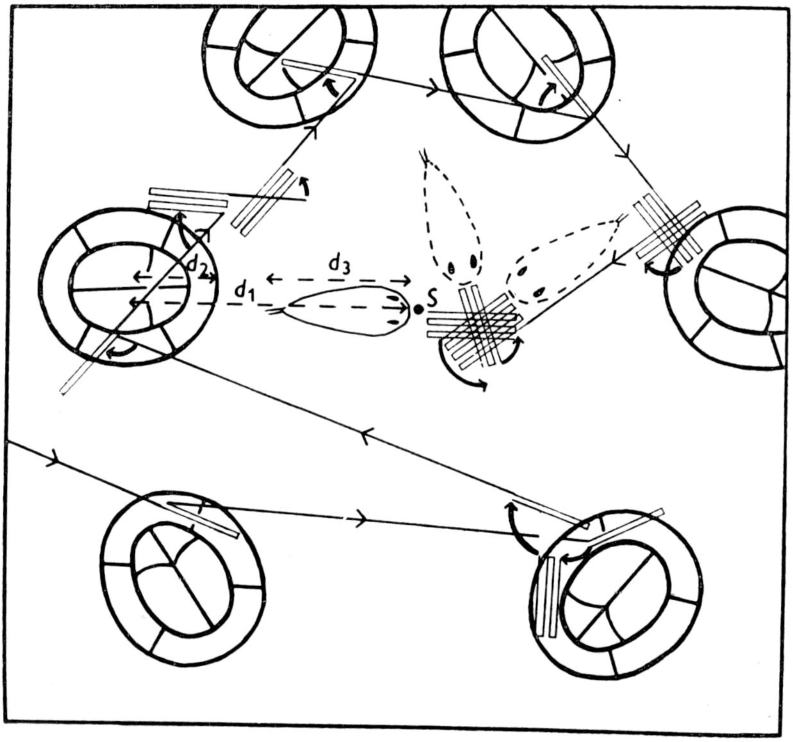

Fig. 1(d)

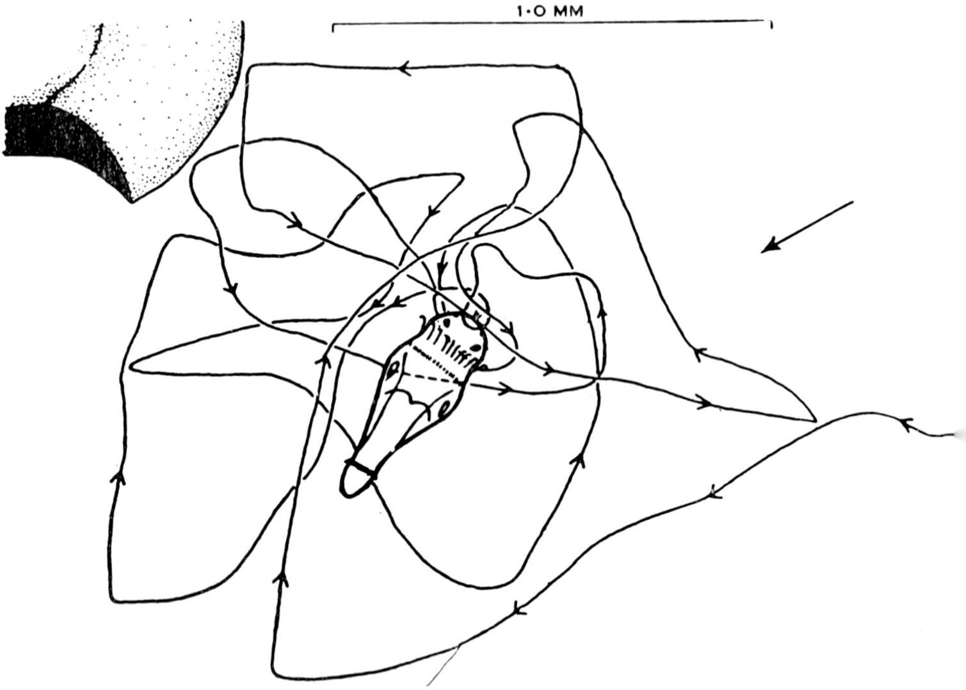

Fig. 1(e)

Fig. 1 Tracks observed during settlement for (a) the barnacle (*Balanus balanoides*, (b) the tubeworm *Spirorbis spirorbis*, (c) the oyster *Ostrea edulis*. Details of the final part of the settlement pattern in which obstacles are avoided is shown for *Balanus balanoides* in (d) and for *Spirorbis* in (e) Magnification of figures: (a), ×4; (b), ×8 (see scale); (c), ×5 (scale drawn = 5 mm); (d), ×20; (e), ×60 (see scale). (Data from Cranfield, 1971; Crisp, 1961, reprinted with permission of the Journal of Experimental Biology; Crisp and Austin, 1960, reprinted with permission of The Annals of Applied Biology; Wisely, 1960, reprinted with permission of The Australian Journal of Marine and Freshwater Research.)

ultimately the animal is restricted to a to-and-fro movement pivoted at a single spot (Knight-Jones and Crisp, 1953; Crisp, 1961). But in bivalves such as *Teredo* (Isham and Tierney, 1953) and oysters (Prytherch, 1934; Yokota, 1936; Cranfield, 1971) the *modus operandi* changes abruptly. The initial fast and straight gliding movements depend on the action of the pedal cilia, while the slower and more deliberately exploratory movements are made by expanding and contracting the foot, aided by secretions of groups of pedal glands which exude threads for temporary attachment. The tip of the foot serves as a probe to inspect the substratum and its activity increases as the larva becomes more excited. The larval movements of *Spirorbis* have not been as fully analysed. *Spirorbis* larvae too have been observed to secrete adhesive threads which prevent their becoming detached by water currents as they glide on their ventral cilia (Knight-Jones, 1951; Gee and Knight-Jones, 1962). Towards the latter end of their exploratory path they change direction violently by lateral flexure of the body. *Bugula neritina* larvae similarly produce adhesive threads (Lynch, 1947).

In addition to the similarities between the behaviour of one kind of larva and another, there are close analogies between larval settlement as a whole and other forms of instinctive behaviour. Using the language of ethology, the settlement "drive" expresses itself in "appetitive" or searching activity which continues until the animal is presented with a specific releasing mechanism leading to the next stage in the behaviour pattern. The logical conclusion to the succession of appropriate releasers is fixation and metamorphosis, representing the "consummatory act". Several observers have noted that larvae may show "overflow activity" if they are kept for a long period with the specific releaser denied. The drive to settle becomes irresistible and metamorphosis takes place in an unsuitable place: for example, in mid-water or on the surface film. Lowered discrimination, abnormal development or death may also result as described above.

A most important question in the study of settlement behaviour is whether larvae automatically occupy suitable sites or whether they exercise a genuine power of choice. Many observers have noticed that if larvae are offered a number of possible sites for settlement, they will usually settle first on those with certain properties which, for brevity and without implying action at a distance, can be referred to as the most attractive. After these sites are filled, later settlers will fill up the remaining sites and, finally, quite unsuitable locations will become occupied. It seems probable that the larva must be capable of choosing in the sense that the first to settle would probably have explored and rejected the less attractive sites before it found and settled on the more attractive. Meadows and I have proved this was so by offering cyprids pits of a dimension which they preferred, together with grooves surrounding the pits which, though somewhat attractive, were less so than the pits (Fig. 2). It was observed that the grooves, though much greater in total area, were never occupied before the pits. Some of the pits were surrounded by one, and others by two, concentric grooves, but this had no effect on

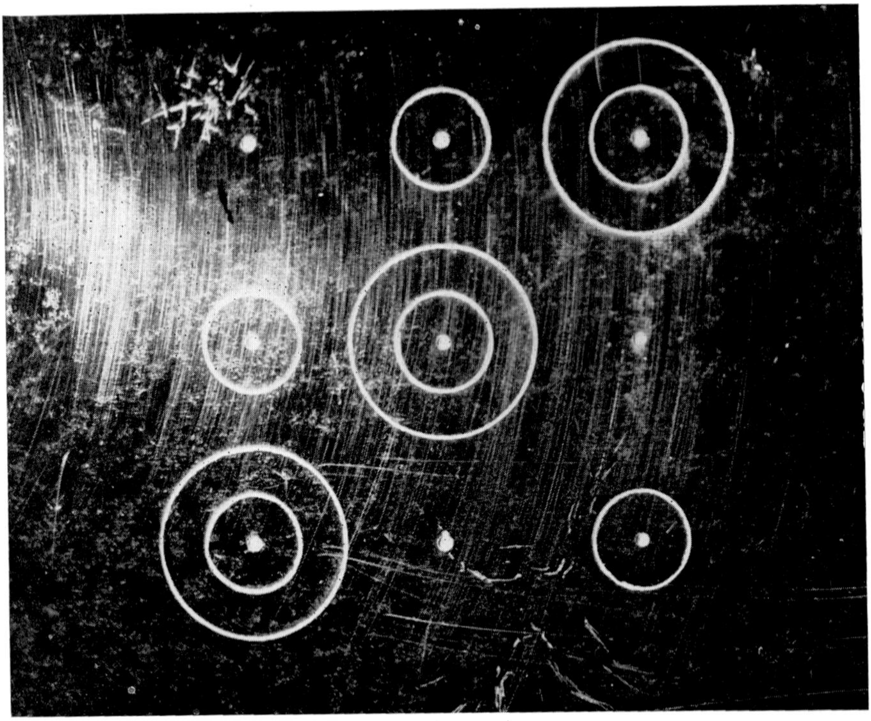

Fig. 2 Experimental design to demonstrate choice of site by cyprids of *Balanus balanoides*. The central pits are of the most favourable dimensions to stimulate the rugophilic response, the surrounding grooves are less favourable. It will be seen that each pit has been filled by single individuals before any other site is occupied. The behaviour pattern of the cyprid illustrated in Fig. 1(a) would have ensured that the grooves would have been entered and rejected before these individuals could have found the pits. Subsequently, when exposed to further settlement, the grooves were occupied.

their occupation. Evidently the cyprids that were first to settle and had the advantage of entering the pits must have prospected and rejected the less attractive grooves before doing so.

The apparently purposeful choice can be explained mechanistically on the basis of a gradual lowering of the threshold stimulus necessary to release the "consummatory act" as the animal becomes older, or as it is stimulated by more prolonged exploration of the surface. The first colonizers with a threshold high enough to be stimulated only by the most attractive sites will occupy them first. Later arrivals will then not be able to settle until their threshold is lowered and brought within the compass of the less attractive sites, and so on. The test of this theory would be to offer a standard stimulus to cyprids of increasing age and, if the hypothesis were correct, an ever-increasing proportion should settle. This was indeed shown to be the case (Fig. 3). After an abnormally long delay, however, the viability of the larvae may diminish and settlement become impossible, as described above.

VI. Responses of Larvae at Settlement

Among the factors of importance to the choice of site by the settling organism are the following: the texture, contour and permanence of the

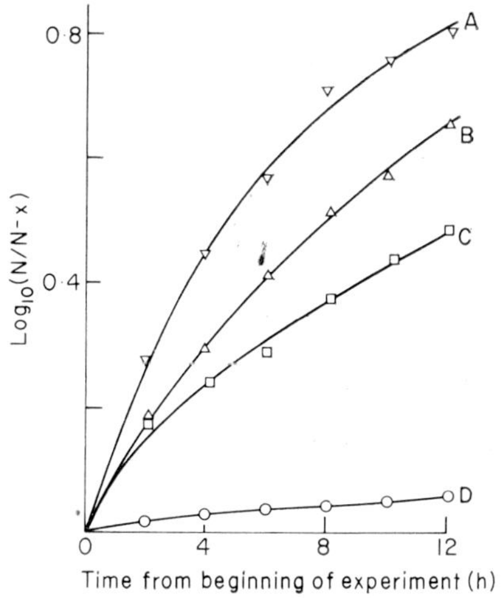

Fig. 3 Rate of settlement of cyprids after having been kept for increasing periods in glass dishes in the laboratory. The settlement rate has been adjusted to allow for the numbers left in the dish (see Fig. 4). The individuals kept longest settle most rapidly: A 5·5 days, B 3·5 days, C 1·5 days and D 1·0h after collection from the sea. (From Crisp and Meadows, 1963. Reprinted with permission of the Royal Society.)

surface; its likely exposure to water flow, light, extremes of temperature and salinity; the availability of food; the possible danger from predators and competitors and the chances of success in reproduction. During its exploration, the larva needs to be able to react to whatever clues it may

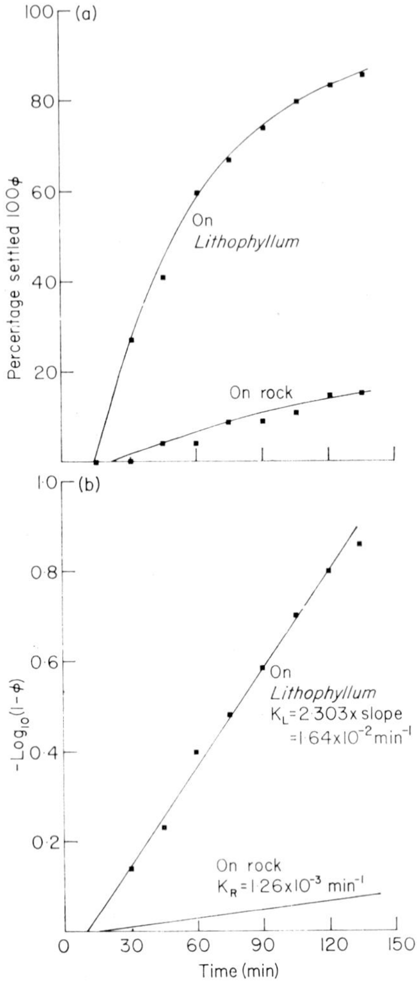

Fig. 4 Form of presentation of data on rates of settlement. The illustration is based on the settlement of *Spirorbis rupestris* larvae on surfaces bearing *Lithophyllum* and on bare rocks of similar size (Gee and Knight-Jones, 1962). The effect of the transformation from $\phi = x/N$ (where ϕ is the proportion settled, x the number settled and N the total number originally present in the dish) to $-\log_{10}(1-\phi)$ is to make the dependence on time linear if the intrinsic rate of settlement remains constant. A. Plot of ϕ against time. B. Plot of $-\log_{10}(1-\phi)$ against time. In Fig. 3, unlike the example here, there was a genuine drop in the intrinsic rate of settlement of cyprids, presumably because the population was heterogeneous and the cyprids more ready to settle did so at a greater rate.

encounter that will indicate how far the habitat will meet these requirements. It must react to favourable clues by entering the next stage in the settlement pattern and to unfavourable clues by swimming off and testing another site elsewhere.

Many experiments have been performed to identify the factors to which larvae react. Usually, the experimenter offers a choice to a group of unmetamorphosed larvae either in the field or in the laboratory and infers their behaviour from the numbers that subsequently settle and metamorphose. More rarely, the observer watches the process and thereby obtains direct evidence on the rate of settlement and how the choice is made. When measuring rates of settlement in experimental containers, it is important to remember that as the larvae settle fewer are left in the pelagic stage and the rate of settlement with time will appear to fall off (Fig. 3). If the total number of settling larvae (N) is known, it is quite easy to allow for the number that have already settled so that a steady rate of settlement will give a straight line. Instead of plotting the number metamorphosed (x) against time, the function log $(N/N-x)$ should be plotted or, which amounts to the same thing, the function $-\log(1-\phi)$, where ϕ is the proportion metamorphosed x/N. An example taken from the work of Gee and Knight-Jones showing the linearization of such data is shown in Fig. 4. The slope $-d\log_{10}(1-\phi)/dt$ gives the rate of settlement in units of reciprocal time.

A. Responses to Mechanical Contact with a Surface

It is unusual for a sessile organism to fix permanently immediately it touches a surface, though possibly some algal spores and bacteria adhere automatically on making contact. The details of the adhesion of the zoospores of *Enteromorpha* have been described by Evans and Christie (1970). An adhesive is present in the form of electron-opaque vesicles situated at the front end of the swimming zoospore. These are lost when the zoospore comes into contact with a surface, and a fibrillar substance, believed to be derived from them, appears to hold the zoospore on to the surface. Proteases will digest both the intracellular vesicles and the secreted substance; α-amylases do not digest either, but reduce the adhesion of the zoospore to the surface. The implication is that the fibrillar substance is a protein-based mucopolysaccharide which serves as an adhesive (Fig. 5). Ascidian tadpoles, towards the end of their free swimming phase, secrete a viscid substance from glandular cells of the adhesive papillae, which result in the larvae sticking to any surface with which its anterior end comes into contact (Grave, 1920, 1944). Grave described how the settled larvae of *Amaroucium* appear to struggle to free themselves, clearly suggesting that attachment is

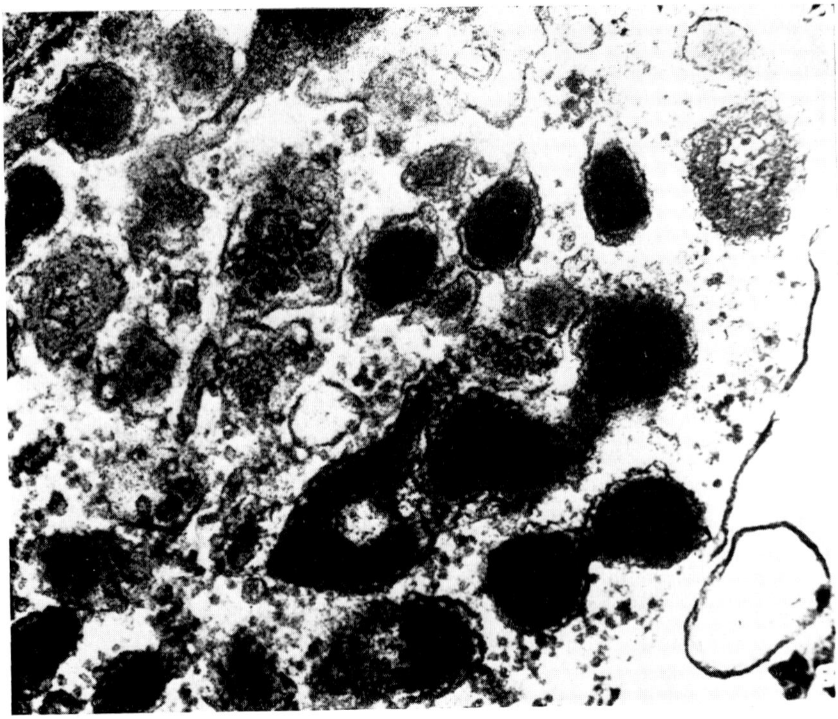

Fig. 5(a)

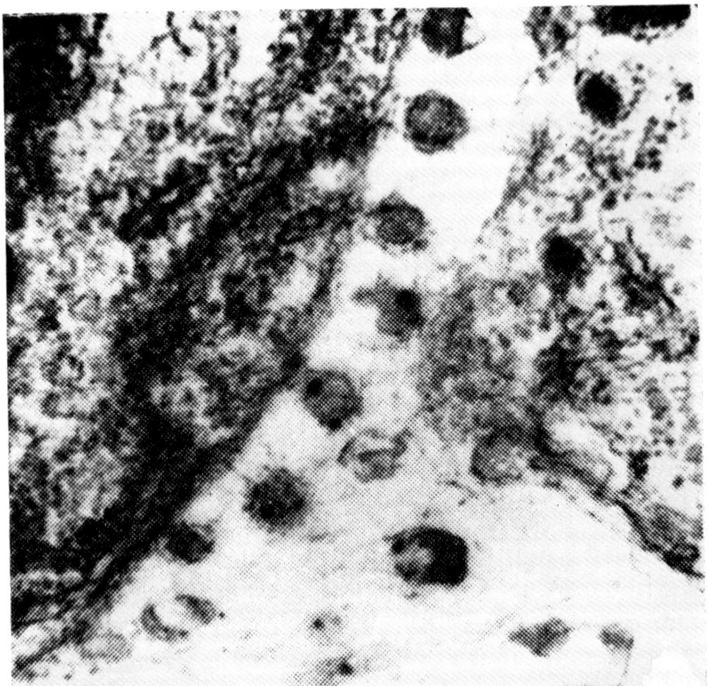

Fig. 5(b)

involuntary and accidental. Later, however, after mentioning the difficulty experienced in trying to dislodge adherent larvae of the *Botryllus* type with a water jet, he states that voluntary release is possible as soon as the larva resumes swimming (Grave, 1932). These larvae can evidently attach temporarily and can be seen to do so as they swim along the blades of eel grass. There are three conical papillae which form a triangular cup-like cavity believed to act as a temporary holdfast

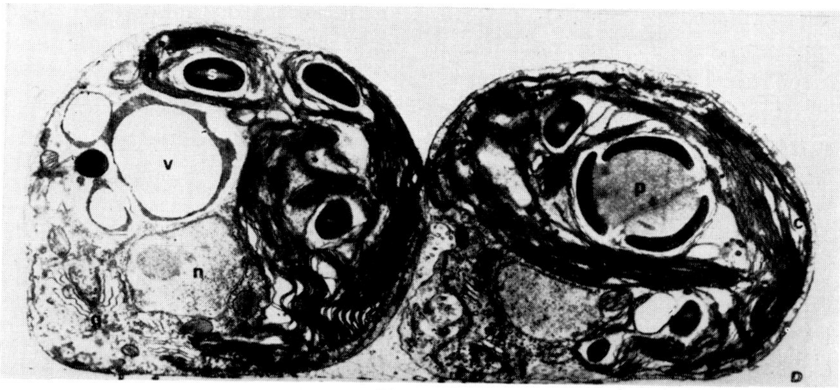

Fig. 5(c)

Fig. 5 Fine structure of zoospores of *Enteromorpha intestinalis* illustrating the discharge of vesicles believed to contain the adhesive which enables the spores to attach on contact with a solid surface. (a) Section through anterior region of swimming zoospore showing electron dense vesicles surrounded by membranes with an undulating border contained within the cell. (b) Section of part of a settled zoospore 5–10 minutes after attachment with similar electron dense bodies lying in fibrillar material outside the cell membrane. (c) Section through two settled zoospores showing fibrillar material in the triangle formed between the cells and the substratum (which is just below the bottom of the figure). The electron dense vesicles have disappeared. (From Evans and Christie, 1970.) (Reprinted with permission of The Clarendon Press, Oxford.)

when pressed against a smooth surface. Whether ascidian larvae explore by moving across a surface seems very doubtful but the larvae of *Diplosoma* can certainly select the darker areas with considerable accuracy (Crisp and Ghobashy, 1971). Some coelenterate larvae also are believed to settle as a result of simple contact, adhering by discharging sticky threads from nematocysts. Hawes (1958) describes the more or less aimless movements of the Actinula larva of *Tubularia larynx*. The tips of the tentacles discharge adherent nematocysts with increasing readiness as the larva ages, and usually it becomes fixed close to the colony from which it arose. A more interesting example of adhesion on contact is afforded by the planulae of *Hydractinia echinata*, a colonial hydroid growing on the outside of disused gastropod shells inhabited by hermit crabs. The planula stands vertically poised on the sand until a shell is dragged past: the movement stimulates it to stick to the passing shell (Schijfsma, 1935). By this means, the larvae are very likely to find their normal host. It may be assumed in all the above examples and in the absence of other evidence that mechanical contact is a sufficient stimulus. However, other properties of the surface may also play a part: for example, it is hard to explain the differential attachment of algal sporelings of *Polysiphonia lanosa* to *Ascophyllum* on the basis of mechanical factors alone. Even the lowly bacteria appear sometimes to be capable of exploratory movements and selection after contact with a surface; they will detach from unsuitable surfaces and adhere to suitable ones (Sieburth, 1968; Floodgate 1972).

B. Responses to Water Currents

It would appear most unlikely that any larva would be able to appreciate water movements while it is being carried freely in a current, since without visual reference points it would have no means of knowing its velocity. No experiments have been done, or thought worthwhile doing, to test this possibility. But it is quite conceivable that changes in light intensity or other optical clues resulting from its passage past objects might evoke a response which would bring the larva into contact with a surface, for example causing it to settle in the shade of a dark object.

When the larva carried in a current is brushed against a surface an entirely different situation arises. Indeed, it is not easy to distinguish between the chance collision of a swimming larva with a stationary object and a collision resulting from a larva being carried against a surface by a stream of water. Either would be likely to evoke a contact stimulus which might result in adhesion as described in the preceding section. But the relative velocity involved in the second kind of collision might be too great to allow settlement to take place.

Many experiments have been carried out to answer the question "at what current speed can a larva settle?" (Smith, 1946; Doochin and Smith, 1951; Wood, 1955). This question is based on the observation that ships become fouled when at anchor in harbours but not when under way on the high seas (Visscher, 1928a). Unfortunately, it is the wrong question. The flow rate which influences the settling larva is that which is responsible for dragging it along when it is on the point of settling. This velocity is much less than the velocity of the main stream of fluid some distance away on account of the drag exerted by the moving surface on the fluid close to it. Furthermore, the flow near the surface is not related in a simple manner to the velocity of the main body of the fluid; it will depend on the topography of the whole system and the distance from which the larva can attach to the solid. Very close to the surface of the solid the velocity of flow v usually increases linearly with the distance x from the surface, so that at a point x = d, the distance for example at which the larva touches the surface:—

$$v/d \rightarrow (dv/dx)$$

(dv/dx) being known as the velocity gradient. It is the velocity gradient close to the moving object that really defines the conditions that will affect settlement. For certain simple experimental conditions it is possible to calculate the velocity gradient and hence the ability to settle can be related appropriately to the conditions of flow at the surface.

Experiments of this kind have been confined to barnacle cyprids (Crisp, 1955). When coming in contact with a surface, the cyprid is thought to swim against the current, thereby making attachment easier. Above a certain velocity gradient, corresponding roughly to the maximum swimming speed of a cyprid at its closest approach, attachment becomes increasingly difficult. The vertical axis of Fig. 6 indicates success in attachment, the horizontal the velocity gradient. It will be seen that the success diminishes, as expected, at high velocity gradients. But there is an unexpected reduction in the percentage of attachment at low velocity gradients also. It appears that when there is no relative movement of the water, the cyprids are not stimulated to settle. Thus, if cyprids are drawn into a pipette, they immediately attach to the sides, but when the relative movement stops they swim off the surface. Although cyprids may not be able to settle above a certain velocity gradient, once attached it is almost impossible to sweep them off by increasing the current, even to values greatly in excess of those which normally prevent settlement.

Experiments in which sea water containing the larvae of fouling organisms is passed under turbulent conditions through tubes of different diameter or in which a disc is rotated can give useful information on the relative effectiveness of water currents in reducing the settlement of various organisms. Smith (1946) found that at Miami the relative tolerance to currents of the larvae of *Balanus eburneus*, *B. amphitrite* and *B. improvisus* ascended in that order; the results of Doochin, and Smith (1951) in a different apparatus more or less confirmed the above order though, not surprisingly, the absolute limiting velocities were different. Doochin and Smith included the bryozoon *Schizoporella unicornis* which had about the same tolerance to current flow as *Balanus eburneus*, and an unidentified hydroid, the tolerance of which was at least equal to that of *Balanus amphitrite*. Wood (1955), working in Sydney harbour, found that algae (*Enteromorpha*?) settled on the more rapidly moving parts of a rotating disc at a nominal velocity of between 2·2 and 11·2 knots. *Hydroides norvegica* withstood up to 1·5 knots and barnacles up to 2 knots. The main difficulty with such experiments is that, not being able to relate settlement absolutely to surface

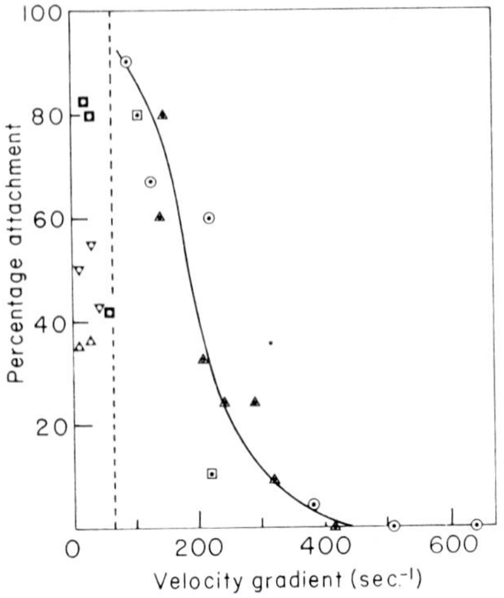

Fig. 6 Experiment demonstrating the influence of velocity gradient on the attachment of cyprids of *Balanus balanoides*. The percentage attaching to the sides of 1 m lengths of glass tubes when water flows through at different rates is given on the ordinate and the velocity gradient at the tube surface on the abscissa. Different symbols denote tubes of different diameter. Below 80 sec^{-1} the numbers fall with decreasing velocity gradient, and their behaviour shows that they lack the necessary contact stimulus at these velocities. Above 100 sec^{-1} the percentage attaching falls as the chance of being able to attach diminishes with higher current speeds. (From Crisp, 1955.) (Reprinted with permission of The Journal of Experimental Biology.)

shear, the results by different workers cannot be satisfactorily compared. But it is clear that a smooth design of ships' hulls or of power station conduits, avoiding regions of low velocity gradient, would help to prevent fouling as long as the fluid is kept in relative motion (Board and Collings, 1965). Many organisms orientate to water currents during or soon after settlement, and the direction that they assume is usually of adaptive significance. Barnacles orientate so that feeding with the cirri is made easier (Moore, 1933; Crisp, 1953; Crisp and Stubbings, 1957), while oysters and tubeworms grow so that there is less chance of their becoming silted up (Board and Collings, 1965).

C. Responses to Surface Texture, Surface Contour and Surface Films

These three features all have some influence on surface topography to which larvae are often very sensitive. The "texture" or "roughness" of a surface is a largely subjective judgement hardly ever defined by investigators. As the scale of roughness increases, it passes imperceptibly into what might better be termed "surface contour". Surface films, which form rapidly as a result of the activity of micro-organisms or by adsorption of large molecules, may influence settlement through their effect on the micro roughness of the surface. They also change its firmness and its suitability as a base for adhesion, while chemical responses to materials present in the film are also possible. Few authors have sought to disentangle these effects.

1. *Surface Texture*

In general, marine invertebrate larvae settle more readily on rough rather than on smooth surfaces. Barnacle species so far tested choose rough rather than smooth surfaces (Barnes, 1956) and are particularly averse to settling on smooth glass, perhaps on account of the diffusion of alkali from its surface. Larvae of *Hydroides* also prefer rough surfaces (Pomerat and Weiss, 1946), but this is by no means universal. Actinula larvae of *Tubularia larynx* were indifferent to roughness (Pyefinch and Downing, 1949) as were those of *Bugula flabellata*, but *Spirorbis borealis* larvae settled most readily on a smooth surface that had been abraded just enough to render it optically dull (Crisp and Ryland, 1960). The larvae of *Crassostrea virginica* preferred the smooth internal surface of oyster shells to the rough outer surface (Crisp, 1967).

An imperfectly understood influence of the living substratum on the settlement of bryozoa may conveniently be included under the heading of surface texture. Ryland and Stebbing (1971) found that the first zooid (the ancestrula) of *Membranipora membranacea* to settle on the plants of *Fucus serratus* tends to face towards the base of the frond, that is, its orientation is proximal. The ancestrula of *Electra pilosa*, which grows on the same weed, orientates distally at settlement, as does that of *Scrupocellaria reptans* growing on the fibrous bryozoan *Flustra foliacea*. The behaviour is evidently dependent on the particular organism providing support or on the conditions under which it grows, since *Membranipora membranacea* shows no detectable orientation when growing on the kelp *Laminaria hyperborea*. Ryland and Stebbing suggest that orientation may result from a response to water flow over the frond, since in a current the frond will stream out horizontally, water flowing from base to tip. Not all the examples given are necessarily consistent with this explanation; light may be a factor in the orientation of *S. reptans* since *Flustra* colonies grow stiffly and bushily away from the substratum and their direction of growth will roughly correspond with the light gradient. Besides, related larvae (*Bugula neritina*) are known to orientate to light (Lynch, 1947). The third possibility is that the larva can detect a polarity in the surface texture of the frond. This is not unlikely since there is no doubt that bryozoan larvae can recognise differences between old and new fronds (Ryland, 1959; Ryland and Stebbing, 1971; Stebbing, 1971), nor is there any doubt that the subsequent growth of the colony is strongly orientated in relation to the direction of growth of the substratum, though not necessarily in correspondence with the orientation of the ancestrula. These difficulties in interpretation indicate the limitations inherent in field observations not supported by laboratory experiments.

2. *Surface Contour and Territorial Behaviour*

When the surface offered has irregularities much larger than the dimensions of the larva, reference may be made to its contour rather than to its texture. Many, if not the majority of lavae, can locate and settle in concavities, no doubt on account of the advantage of shelter which they offer. Boring organisms, such as *Xylotrya gouldi*, will seek a depression as a likely point of attack on the substratum (Sigerfoos, 1907). Barnacle cyprids not only select and settle more densely in pits with small radii of curvature (Crisp, 1961) but they also orientate in the direction of narrow grooves so that they receive the maximum protection. Moreover, very shallow grooves or pits with radii of curvature much larger than the dimensions of the cyprid can be perceived and the cyprids will settle in greatest abundance at the bottom of such concavities. Crisp and Barnes (1954) termed such behaviour "rugotropism". In barnacles, rugotropism seems to be mediated by tactile rather than visual response, since in the dark or with a light shone normally to the plane of the surface, most of the cyprids will still settle in the grooves. Ryland (1959) found that the larvae of several polyzoa settled in concavities. *Celleporella hyalina* larvae chose concavities on the fronds of *Laminaria saccharina* which were extremely large in comparison with the dimensions of the larvae themselves. He argued that neither gravity, light nor surface texture could easily explain the preference and siting, and that the larva during its gliding rotatory path just before settlement must be able to detect the curvature of the surface. The cause of rugotropism is apparently different in the larva of the ascidian *Diplosoma listerianum* in being dependent, not on tactile responses, but on the shadow cast within the groove by its walls (Ghobashy, 1967).

Rugotropism is widespread not only throughout the animal kingdom but is displayed also by the motile spores of certain plants. Müller (1964) demonstrated quantitatively that *Ectocarpus* spores settle in grooves on a Teflon surface, avoiding the plane areas adjacent. Christie (private communication) has found that the zoospores, though not the gametes, of *Enteromorpha* search out depressions and aggregate in them at settlement with perhaps an even more pronounced affinity than shown by *Ectocarpus*.

Surface projections usually have the opposite effect, sometimes causing the larvae to swim off, or at other times to avoid the obstacle and settle at a short distance from it. As a result of this behaviour, surfaces with a fibrous texture may discourage settlement. Barnes and Powell (1950) immersed in the sea staple cloths which had fine projecting filaments 8μ in diameter made from glass fibre. These surfaces, when exposed in the sea, were avoided by barnacle and tubeworm larvae, though not by the larvae of *Tubularia*. Surfaces covered with mats of algal filaments are similarly avoided by most animal larvae. However, an exception is to be found in the pediveliger larva of *Mytilus edulis* which attaches to algal filaments of about 180μ diameter and which remains on them during a temporary pre-settlement phase (Bayne, 1964b). Since fibrous collectors are commonly used to encourage the settlement of *Pecten* larvae in Japan and in the U.S.S.R. (Golikov, A. N. private communication) it is possible that a number of bivalves show similar behaviour. The fibres evidently afford a suitable temporary habitat where the larva has access to water currents and can grow rapidly without the disadvantages of competition and siltation that it might encounter in the adult mussel bed.

Swimming off or settling at a distance as a result of an encounter with an obstacle projecting from the surface may have an important function in preventing the species from populating a habitat too densely. If the settlement can be spaced out evenly, growth and reproduction are likely to be maximized. Wisely (1960) found that the larvae of *Spirorbis borealis* were attracted to settle in the linear grooves formed by the "midrib" of *Fucus serratus* but their distribution was not random. They were spaced apart in a significantly regular manner. They spaced themselves out similarly on a plane surface. When just about to settle, the larva prospects the chosen area in a roughly spiral path and, should it meet an obstacle, whether a sand grain or another individual, it turns and prospects another area a little way off (Fig. 1e). In this way, it is likely to avoid any established individual within a distance very roughly equal to the diameter of the final prospecting path. Wisely found that if large numbers of larvae were forced to settle on a limited area of substratum, the pre-settlement behaviour eventually became abbreviated and the prospective path was reduced to a few turns with the result that the larvae then settled abnormally close together.

Such behaviour, though essentially a tactile response, has close analogies with territoriality in higher animals. First, the more recently settled individuals seem to respect the territory of the established ones. Secondly, the size of the territory diminishes with increasing population pressure. Territorial behaviour in this sense is shown not only by *Spirorbis* but also by barnacle cyprids (Fig. 1d, Crisp, 1961). Knight-Jones and Moyse (1961) put forward evidence that barnacle cyprids recognized and avoided their own species but would settle in contact with other species. They argued that this behaviour would be beneficial to its own species by reducing intraspecific competition, and at the same time be harmful to alien species. The experiments were criticized by Crisp (1961) on the grounds that differences in the shapes of the margins of established individuals of the different species might influence settlement. Before it can be accepted that chemical recognition plays a part in territorial behaviour, further work is needed in which the possible effects of the size and shape of established individuals has been eliminated.

3. *Particle Size of Deposits*

Larvae of burrowing species no less than those of animals living on exposed surfaces need to choose the right habitat. They investigate substrata tentatively, only metamorphosing when they have entered a deposit with suitable characteristics. The obvious factor which greatly influences the infaunal environment is particle size. Wilson pioneered investigations which have shown the importance of the grade of the deposit in accelerating metamorphosis of polychaete larvae; for example *Owenia fusiformis* (Wilson, 1932), *Notomastus lateraceus* (Wilson, 1937) and *Scolecolepis fuliginosa* (Day and Wilson, 1934). At the outset of his classical investigations on the larvae of *Ophelia bicornis* Savigny, he inclined to the view that the dimensions of the interstitial voids between sand grains was the chief factor causing metamorphosis. Larvae did not metamorphose readily in sands where the voids were so large that the animal did not make sufficient contact with the grains; sands of a grain size too small to allow penetration were likewise rejected. Similar results were obtained using artificial sand made from glass beads. Evidently, the larvae of *Ophelia* at metamorphosis behave similarly to most interstitial organisms in selecting voids of the appropriate size, presumably by thigmotaxy. Gray (1966a, 1967a) demonstrated that the strongly thigmotactic interstitial worms *Protodrilus symbioticus* (Giard) and *Protodrilus hyperleucus* Amanante choose sands with a modal grain size of about 300μ. Although these were adult archiannelids, their behaviour is relevant not only because they are of similar size to invertebrate larvae but because the group to which they belong is probably neotenous. But, as will be discussed below, particle size is only one of the factors inducing metamorphosis in burrowers.

4. *Surface Films*

Within minutes of being placed in natural sea water, an inert surface is likely to adsorb high molecular weight substances and become colonized by microorganisms which secrete hydrophilic colloids commonly known as "slime films" or "surface films" (Zobell and Allen, 1935; Zobell, 1939). Though Hilen (1923) had demonstrated otherwise, there exists in the literature a number of unpublished reports which, confusing *propter hoc* with *post hoc*, maintain that a slime film is a prerequisite for the settlement of larger organisms. This mistaken view unfortunately had considerable influence but was refuted by the work of Miller *et al.* (1948) who showed conclusively that *Bugula neritina* sometimes settle more readily on surfaces from which the film had been wiped off than on a filmed surface. Crisp and Ryland (1960) showed that larvae of *Bugula flabellata* actually preferred unfilmed surfaces. So also do those of *Alcyonidium polyoum* (Crisp and Williams, 1960). A particularly interesting example of preference for a non-filmed surface is afforded by the larva of *Hydractinia echinata* (Teitelbaum, 1966). It needs to select the surface of a shell occupied by a hermit crab and such a shell would be constantly abraded through being dragged over the sand.

Nevertheless, it is true that the majority of larvae studied so far settle more readily on filmed surfaces. Examples can be drawn from barnacles and tubeworms (Daniel, 1955), oysters (Cole and Knight-Jones, 1949), *Spirorbis* (Knight-Jones, 1951) and the cosmopolitan tropical bryozoan *Watersipora cucullata* (Wisely, 1958a). Few investigators have examined the responses of larvae to surface films derived from different organisms or in different states of metabolic activity. Ryland (1959) held that the mucus secreted by brown algae was a deterrent to the Bryozoa that normally settle on *Fucus* fronds. Meadows and Williams (1963) reported that substances formed by the microalga *Dunaniella* discouraged the settlement of *Spirorbis* but films formed by *Navicula* and mixed diatoms encouraged settlement unless they had been kept in the dark. A similar phenomenon was reported by Knight-Jones *et al.* (1971) who found that *Fucus* fronds were much less favourable to this species after having been kept in darkness or if their metabolism was inactivated in other ways.

Perhaps the clearest influence of the great importance of surface films is to be found not so much in the context of substrata exposed directly to the sea but rather in connection with films on the intimate surface of sand grains. It is now known that many invertebrate larvae of burrowing forms, as well as adults of interstitial species, can respond to the presence of suitable deposits after making contact with the particles of which they are composed. In 1940, Jägersten found that the property associated with deposits which induced the metamorphosis of larvae of *Protodrilus rubropharyngeus* could be transferred to shell gravels and small stones to which the larvae had previously shown no response. Jägersten's subsequent experiments were not entirely consistent with the conclusion that he reached and which Gray (1967a, b) subsequently demonstrated, namely, that the attractive factor was of bacterial origin and could be transferred through simple infection of the surface of the gravel. It was D. P. Wilson, (1954, 1955) who first clearly demonstrated the importance to marine benthic larvae of the microflora present on the surface of sand grains. His earlier papers (Wilson 1952) showed that *Ophelia bicornis* larvae were very sensitive to the origin of the sands. Sand from its natural habitat was the most successful in promoting metamorphosis, but sand from a nearby locality was inhibitory. A most important series of observations followed, establishing that alterations in the properties of the surface alone profoundly modified the attractiveness of the sand. After heating to redness, attractive and unattractive sands were found to approximate in quality. However, re-immersing such heated sands in sea water restored their attractiveness. Wilson concluded that a surface microflora of a specific type and/or density was required. Gray (1966b, c, and 1967a, b) carried out a series of similar experiments using adults of *Protodrilus symbioticus* and *Protodrilus hypoleucus* and the larvae and adults of *Protodrilus rubropharyngeus*. He found not only that highly destructive agents would permanently destroy attractiveness but also that quite subtle treatments of the sand, such as drying out in air (Gray, 1966c), would render them temporarily unattractive to the archiannelids. He thus confirmed that these organisms, like *Ophelia*, required a delicate organic film probably of microfloral origin, on the surface of the sand. He then tested a number of bacteria for their capacity to re-establish the attractive quality in sands from which the natural factor had been removed by destructive agents. Cultures of naturally occurring sand bacteria were found to be the most effective and restored the sand to its original condition. Other bacteria, such as *Pseudomonas* sp., and terrestrial soil bacteria, which normally grow attached to surfaces, increased the attractiveness appreciably, whereas non-sessile bacteria increased it to a lesser extent. By maintaining inert sands for different periods in an enrichment culture, and then shaking off the bacteria, Gray was able to alter the density of bacterial cells remaining at the sand surface. He found that the attractive quality did not vary *pari passu* with the quantity of microfloral organisms present. The natural sands were usually more attractive than artificially restored sands though they were much less heavily coated with bateria. Gray concluded that an adherent film produced by the bacteria, rather than the bacteria themselves, was probably the factor searched for by the larvae. If this is true, it follows that the larvae are primarily searching, not for a food source, but for a habitat indicator. An experiment in which larvae were offered sands which had been treated with cultures of microalgae, which constitute their main diet, was particularly instructive. One was treated with an infected unialgal culture, the other bacteria-free. The larvae showed a strong preference for the former though the difference in food content would have been negligible (Gray, 1966b).

The mechanism by which the surface property of the sand is recognized is not known. The capacity appears to be common to a number of interstitial organisms (Gray, 1968; Gray and Johnson, 1970) as well as to many benthic larvae. Both Wilson and Gray, who have repeatedly observed the reactions of the organisms to sand grains, make the point that they do not respond until they have made actual contact, indicating that close range physical forces rather than diffusing chemical substances are responsible for the attractiveness.

From all the foregoing observations on surface films, whether on exposed or interstitial surfaces, it is clear that the influence they have on settlement is not a simple physical effect due to the presence of a viscous or adhesive colloid. Physico-chemical or chemical factors must be involved and these will be discussed later.

D. Responses to Light and Gravity

Since in the normal environment with overhead illumination a positive response to light and a negative response to gravity offer alternative means of orientating to the vertical, they are conveniently considered together. Simple responses to light by swimming larvae are immediately obvious to the observer and have been reported so frequently that they would be impossible to list. Responses to gravity need to be demonstrated experimentally and are therefore less often reported though probably quite as common. Responses of the larvae to light and gravity are much less evident at settlement than when the larvae are in the swimming-stage and the literature is correspondingly ill-supplied with reliable information. Responses to light and gravity are often inferred from the pattern of distribution of settled larvae on the surfaces of plates exposed in the sea, but such interpretations may be uncertain unless great care has been taken in designing the experiment. An

Table 2. Responses to light and gravity of a selection of marine invertebrate larvae

Group and Species	Author(s)	Reaction during searching		Reaction at or near settlement		Position occupied		Remarks
		Gravity	Light	Gravity	Light	Surface Angle	Relation to light	
Sponges								
Microciona prolifera								
Cliona sp.	Warburton, 1966	–	*	+	*			Settling larvae swim downwards
Haliclona sp.	Bergquist and Sinclair, 1968	+	+		+	B, L		Under stones in bright light
Halichondria moorei	Sinclair, 1968	+	*	–	– ?	S	I	
Ophlitaspongia seriata	Fry, 1971	–	*		*		S	In places exposed to water currents
Mycale macilente	Bergquist and Sinclair, 1968		–		–		S	
Polymastia granulosa	Bergquist *et al.*, 1970		*		*	B		
Hydroids								
Clava squamata	Williams, 1965		+		–		S	Response altered in presence of *Ascophyllum nodosum*
Actinians								
Lebrunia sp.	Duerden, 1899					L	S	
Siderastria radians	Duerden, 1905	–		+		B		Settle gregariously
Dendrophyllia manni, *Cyphastrea ocellina*, *Pocillopora caspitosa*	Edmondson, 1929	–	+		–	B	S	*Cyphastrea* settles less readily in darkness
Fungia actiniformis	Abe, 1937		+	+	–	B	S	Delayed settlement in dark

Polychaetes								
Salmacina dysteri	Giard, 1876		+		—	S	S	Adults photonegative
Hydroides dianthus	Zelenky, 1905		+	— ?	+	S, U	I	Settle at waterline
H. norvegica	Wisely, 1958b							
H. hexagonus	McDougall, 1943		+		+	S	I	
Pomatoceros triqueter	Segrove, 1944		+	— ?	+	S	I	Settle at waterline
Spirorbis borealis	Knight-Jones, 1951	*	+	*	—	*	*	
	Knight-Jones, *et al.*, 1971							
Sabellaria vulgaris	McDougall, 1943					U	I	Adult can survive sedimentation
Bivalves								
Mytilus edulis	Bayne, 1964a							
Trochophore		*	*					
Young veliger		*	—					
Straight hinge veliger		*	+					
Veliconcha		—	*					
Eyed veliger		—	+					
Swimming pediveliger		?	—	—	—			Anterior end settles away from light
Crawling pediveliger		?		—	*			
Ostrea, *Crassostrea*	Cole and Knight-Jones, 1939, 1949; Hopkins, 1935, 1937; Crisp, 1967		*	—	*	L	S	Settle hinge upwards
Teredo pedicellata	Rancurel, 1951; Isham *et al.*, 1951; Isham and Tierney, 1953	— ?	+	*	—	B	*	Larvae show low photokinesis

Table 2 (*contd.*)

Group and Species	Author(s)	Reaction during searching: Gravity	Reaction during searching: Light	Reaction at or near settlement: Gravity	Reaction at or near settlement: Light	Position occupied: Surface Angle	Position occupied: Relation to light	Remarks
Polyzoa								
Bugula neritina	McDougall, 1943; Lynch, 1947; Ryland, 1960; Mawatari, 1951	—	+	+	+ ?	L	S, L	Settles with eyespot towards light. A few reversed sign of phototaxis. Settles in light and dark places
Bugula turrita (Described as *B. flabellata*)	Grave, B. H. 1930; Lynch, 1949b	—	+		—			
Bugula simplex; *Bugula fulva*; *B. plumosa*; *Scrupocellaria scruposa*; *Scrupocellaria reptans*; *Celleporella hyalina*	Ryland, 1960; Lutaud, 1953; Ryland, 1960		+		—		S	
Cryptosula pallassiana	Ryland, 1962a							
Alcyonidium polyoum	Ryland, 1960		*		*	*	*	
Alcyonidium hirsutum; *Flustrellidra hispida*	Ryland, 1960		+		+			
Watersipora cucullata	Mawatari, 1952; Wisely, 1958a		+	— ?	±	L	S	At settling stage normally photonegative but swam photopositively when in bright light
Barnacles								
Balanus balanoides	Crisp (unpublished); Barnes *et al.*, 1951							
Nauplius St. 1		*	+					
Late Nauplius		*	±					

Barnacles (*contd.*)								
Cyprid				−	+	L	S	Orientate to light. Sink on sudden variation in light intensity
Elminius modestus	Crisp and Ritz, 1973		±	−	+	U	I	Position based on lab. experiments
Balanus amphitrite, *B. eburneus*, *B. improvisus*, *Lepas anatifera*, *Chthamalus fragilis*	Gregg, 1945; Smith, 1948; McDougall, 1943; Skerman, 1958; Visscher, 1928b		+		?	L	S	Visscher mistook direction of orientation. See Forbes *et al.*, 1971
Simple Ascidians								
Ciona	Berrill, 1947; Miller, 1953; Dybern, 1962	−	+	+	−		S	Difficult to distinguish geopositive responses from passive sinking
Styela partita	Grave, 1941, 1944	−	*	−	*	B	S	Eye poorly developed. Swim up with reduction in light
Molgula citrina	Grave, 1926; Crisp (personal observation)	−		−	*	L	S	Eyes absent
Perophora viridis	Grave and McCash, 1923	*	+	*	±		I	Settles in bright light in littoral
Compound Ascidians								
Amaroucium sp.	Grave, 1920; Mast, 1921		+	+	−			
Diplosoma listerianum	Crisp and Ghobashy, 1971	*	+	−	−	L	S	Sink at high light intensities and high temperatures

Key: + Photopositive, or geopositive (sink or swim down); − Photonegative, or geonegative; ± Some larvae give positive, some negative responses; * indifferent.

Relation to light S. Shaded situations I. Well illuminated situations

Surface angle: U. Settle on upper surfaces L. Settle on lower surfaces S. Settle on sides of dish B. Settle on bottom of dish.

attempt has been made in Table 2 to bring together under phylogenetic groupings the more extensive and reliable observations.

There is a great contrast between the predominantly photopositive behaviour of the newly formed or recently released larva and the variable responses of the larva at settlement. Furthermore, the behaviour to light of the settling larvae under laboratory conditions does not always accord in a simple way with its settlement pattern under natural conditions. For example, McDougall (1943) found that the larvae of *Bugula neritina*, though continually positive to light in the laboratory, settled preferentially on the lower parts of pilings which would receive least illumination, on the underside of horizontal plates and in the least illuminated parts of a light box. Similarly, barnacles settle predominantly by day and appear to require artificial light of a minimum intensity to cause significant settlement at night (Weiss, 1947); yet they settle on dark surfaces (Pomerat and Reiner, 1942; McDougall, 1943; Smith, 1948) and on under surfaces shaded from the light (Gregg, 1945). Moreover, the occupation of a particular habitat does not necessarily imply a similar response to light on the part of the larva. For instance, the responses of the larvae of species epiphytic on *Fucus serratus* in the lower littoral zone show all degrees of reversal of phototaxis at settlement; *Flustrellidra hispida*, no reversal; *Alcyonidium hirsutum*, partial reversal; *Spirorbis borealis*, complete reversal; while the fourth, *Alcyonidium polyoum*, is indifferent to light.

An obvious reason for such anomalies is the masking of the influence of light in the field by factors which are carefully excluded in the laboratory—the presence or absence of surface films, the specific attraction of certain substrata, the competition for space by other organisms and the loss of spat through predation, scouring or siltation. However, there remain a number of experiments in which some other explanation must be sought. It is possible that the obvious reactions to light may not be those which are important at the moment of settlement. Despite the clear implication in many field observations of the effects of shadows, most investigators have ignored the possible influence of light intensity and particularly of sudden changes in light intensity. If the drop in illumination on passing into the shade of an object were momentarily to reverse the normal photopositive response or to reduce the threshold of a photonegative response, the larva would be suddenly directed towards and would probably alight on the object. The settlement behaviour of barnacles could be more readily understood if the larvae first aggregated into the zone of optimal illumination under the influence of positive phototactic responses to the direction of the light, while at close range the behaviour was modified by photokinetic responses to sudden changes in the intensity of illumination. A reversal of phototaxis would be one such possibility; another would be a negative response to gravity causing the larva to swim vertically upwards when passing into shadow. Crisp and Ghobashy (1971) claimed that in many ascidian larvae there is experimental evidence of low photokinesis (in the sense of Fraenkel and Gunn, 1940) resulting in upwards swimming, which would explain the pattern of settlement.

Many field experiments dealing with the efficacy of spat collectors have been made to test whether organisms settle best on the upper, the vertical or the lower surfaces of plates, and it has become conventional to refer to these positions as 0°, 90° and 180° surface angle respectively. Clearly, responses to both light and gravity will have an influence on which surface angle is chosen, but so also will the mode of swimming, whether mainly in the vertical or horizontal plane. The relatively small numbers of larvae of most organisms which settle on vertical surfaces suggests that swimming and sinking in a more or less vertical direction is the general rule. Alteration between upwards swimming and downwards sinking probably plays an important part in the behaviour of many groups of larvae and must be borne in mind in accounting for the pattern of settlement. If a horizontal test plate is close to the surface of the sea, there is ample opportunity for larvae swimming upwards to come into contact with its under surface, but little chance of any larvae sinking on to the upper surface. The reverse is true of a plate set near the bottom. Crisp and Ghobashy (1971) found that the numbers of ascidian tadpoles of *Diplosoma* settled on the upper and lower surfaces of a series of test panels at different heights in the water column was strongly influenced by this factor. The transparency of the panels was also important, the disparity in favour of the under sides being much greater with opaque than with transparent panels, presumably on account of their stronger shadowing effect.

It can be seen from Table 2 that the great majority of animal larvae do in fact settle on the under sides and in the shade. Indeed, it is quite exceptional to find organisms, apart from plants, which will select or even tolerate upwards facing surfaces. McDougall (1943) found that *Sabellaria vulgaris* alone among the rich sessile fauna of Beaufort, North Carolina, favoured upper surfaces and thrived there. The heavy siltation of 0° surfaces is harmful to most organisms but is used by this species to construct its tubes which raise it clear of the surface. The larval behaviour of *Sabellaria* would be of particular interest from this point of view and might well help to explain the mechanism of settlement of shade seeking species.

Ascidians and bivalve larvae appear to have well developed responses to gravity. The former possess statocysts, but since at settlement they do not crawl significantly—the colonies themselves may be slightly mobile (Carlisle, 1961)—the statocysts are presumably used for vertical orientation only during the free swimming-stage. In bivalves, however, their negative response to gravity has been demonstrated both when free swimming, and when crawling. The larvae of *Ostrea edulis*, in common with those of other oysters, settle mainly on the under sides of objects. They are remarkable in possessing eyes yet apparently being insensitive to light; indeed Prytherch thought that the eyes were functionless but Nelson claimed to have observed a kinetic response, the larvae ceasing to move only when in shade. Cole and Knight-Jones (1939) confirmed that oysters chose darker surfaces during daylight and, by means of a delightfully simple and conclusive experiment in which a glass funnel was used to concentrate upward swimming larvae, the same authors (1949) showed that under sides were favoured because larvae about to settle swam upwards. A similar tropism continues after settlement, the exploring pediveliger taking a zig-zag upwards path on the under side of plates held at 45° to the horizontal. Pediveligers of *Mytilus edulis* behave similarly to gravity (Bayne, 1964a) though they are significantly negative to light as well, so that in nature they would tend to move upwards and into crevices. Probably the gravity responses in both these species enables the larvae to evade areas where silt deposits are formed. *Teredo* is a third bivalve whose economic importance has stimulated considerable study (Carpine, 1960). The initial photopositive response of the free swimming larva (Rancurel, 1951) soon weakens and the larvae show low photokinesis, dropping by gravity to the bottom when the illumination exceeds 70ft candles. Here they are indifferent to light (Isham, *et al.*, 1951), gliding over the surface in search of cracks and pits. Not unexpectedly, there is no evidence of gravity responses, since the larvae will enter wood at any convenient point.

E. Responses to Chemical Factors

Included among the factors likely to be of importance to larvae in choosing a habitat (p. 203) were a number dependent upon intraspecific and interspecific interactions. Intraspecific competition is to some extent reduced by territorial behaviour which is certainly due in part to a tactile response to protuberances on the surface (page 212). There is a possibility, however, that it could be re-enforced by specific chemical factors. Of much more certain significance in the context of responses to chemical substances is the phenomenon of intra-specific gregariousness, which plays such an essential part in the reproduction and genetic recombination of sessile organisms with dispersed larval stages. To achieve intraspecific gregariousness, a species-specific recognition is necessary at the time of settlement. Incidentally, gregariousness will not only increase the probability of reproduction but also of survival, since it favours settlement in situations where other individuals have persisted. However, for the purpose of survival the gregarious response need not be exclusive; it could equally well encompass a response to other species utilizing similar habitats. As Knight-Jones (1955) has demonstrated, barnacle cyprids can respond to a number of related species and genera.

A slightly different but equally important phenomenon is seen in species which show a preference to settle on other species or higher taxa of quite unrelated plants and animals. The host may merely provide a suitable substratum, it may offer freedom from competition by other organisms, it may be utilized as food or it may serve to indicate a habitat favourable in some other respect. The localization of annelid larvae to appropriate deposits with a particular microflora is an example of such an interspecific association. A very precise set of relationships is afforded by the various species of *Spirorbis* which tend to become segregated at settlement to particular substrata (Table 3). One of these

Table 3. Relationship of *Spirorbis* species to habitats

Species	Favoured habitat
Spirorbis spirorbis L. (= *S. borealis*)	*Fucus serratus*
certain races	*Fucus vesiculosus*
Spirorbis corallinae De Silva and Knight-Jones	*Corallina officianalis*
Spirorbis rupestris Gee and Knight-Jones	*Lithothamnion polymorphum*
Spirorbis infundibulum Harris and Knight-Jones	*Tenarea* reefs
Spirorbis corrugatus Montagu	Blades of *Posidonia*, *Zostera*
Spirorbis inornatus l' Hardy and Quiévreux	*Laminaria* fronds
Spirorbis spirillum (L)	
races in open sea	*Abietinaria abietina*
races in deep lochs, fjords and harbours)	*Laminaria* sp.
Spirorbis armoricanus Saint-Joseph.	*Homarus*, *Palinurus*
Spirorbis pagenstecheri Quatrefages.	Stones, *Laminaria*
Spirorbis tridentatus Levinsen	Rocks, stones and shells
Spirorbis vitreus Fabricius	Rocks, stones and shells

species, *Spirorbis spirorbis*, is known to settle preferentially on fucoids, normally on *F. serratus* (Williams, 1964). A race of this species, living in a recently isolated body of sea water, has been identified experimentally on account of its preference for a different species of *Fucus*, *Fucus vesiculosus* (Knight-Jones *et al.*, 1971). Interspecific relationships between quite different organisms may have evolved with individual survival as a primary advantage, but incidentally, by clustering the individuals of the species into a narrow habitat, the association will lead to an improvement in the chances of reproduction also. If the capacity for substrate selection is inherited, genotypes adapted to particular species of host organisms will become segregated, leading to habitat speciation. The example cited above (Knight-Jones *et al.*, 1971) suggests incipient speciation during a very brief period of less than a hundred years.

Another biotic relationship is manifested by larvae of benthic forms which utilize a particular prey and are stimulated to settle on or near to it. The opisthobranch *Adalaria proxima*, for example, was shown by Thompson (1958) to metamorphose in the presence of colonies of the bryozoan *Electra pilosa* on which it browses. The same author (1964) lists the interspecific relationship of this group of molluscs, most of which have a planktonic stage and will quite probably be found to have specific responses at settlement (*e.g.* Tardy, 1962). Similar relationships might well be expected to exist among the larvae of parasites, but James' (1971) review of experiments with Digenean larvae indicates that host and site selection is surprisingly imprecise and often non-existent.

Whatever the advantage of these intra- and interspecific relationships, all imply a response on the part of the larva to the presence of its own or of other species or groups of species at the time of settlement. In all cases so far investigated, the response appears to be brought about by chemical recognition. I shall employ Knight-Jones' term "gregarious settlement" where the colonizing species is induced to settle by established individuals of its own or of a closely related species. For the other types of relationships, in which the colonizing species is induced to settle by individuals of an unrelated species, I shall use the term "associative settlement".

Evidence for gregarious settlement usually starts with the observation either of clustered individuals or better, of young spat predominantly colonizing an area where adults of the same species are growing. Associative settlement is indicated by the frequent occurrence of the colonizing species A on or in the vicinity of the established species B. However, these patterns are not proof of habitat choice; they could arise from random settlement followed by differential mortality, as was once widely believed, or from the elimination of the dispersive phase as Barnes (1953) suggested for certain barnacles—though in the case of *B.*

balanus and the great majority of other thoracican species he was almost certainly wrong (Crisp and Knight-Jones, 1953). The second piece of evidence, therefore, is the experimental demonstration that the colonizing species settles at a greater rate or in greater density in the presence of the established species than in its absence. In such experiments, the larvae of species A are offered similar test pieces of inert substrata or piles of deposit, some with and some without the established species—species A if it is gregarious settlement that is being investigated or species B if it is associative settlement. Clearly, all factors other than chemical influences should be eliminated as far as possible, including the effect of contour, appearance, colour and water current associated with the established species. Similar test objects with and without small spat are usually presented, either in darkness or at the periphery of a glass dish which is rotated to equalize illumination (Crisp and Meadows, 1963). Ideally, all comparisons between candidate and control test objects should be made in the same dish so that they have equal chances of being colonized. Otherwise, one comes against the difficulty of dispensing larvae equally into a number of dishes and the danger that differences in the settlement potential of different batches of larvae may be present. If it is impossible to accommodate all the variety of objects in one and the same dish of larvae, the experimental design should ensure that comparisons between treatments are not confounded with differences between containers or batches of larvae (Ryland, 1959). The third and most rigorous stage in the demonstration of a chemical stimulus to settlement involves the isolation and transfer of material to an otherwise inert surface to render it attractive. This is really the only way to eliminate completely the possibility that some subtle visual or contour stimulus may be the basis of gregarious or associative settlement. Table 4 lists examples where chemical inducements have been shown or are believed to exist. The three stages of investigation are indicated in column 3 as follows:—S means that gregariousness or association have been suspected from the appearance of settlement patterns observed, usually in the field; D means that substrate selection in respect of the established species has been demonstrated experimentally; while Ex means that an active principle has been extracted from the established species. The symbol B implies that an inert or unattractive deposit has similarly been rendered attractive by inoculation of specific bacteria.

The table shows that settlement inducing factors have been obtained from the following phyla: Coelenterata, Annelida, Bryozoa, Mollusca and Arthropoda; while bacterial flora have been shown to influence settlement in several groups strongly represented in the interstitial environment: Polychaeta, Archiannelida, Gastrotricha and Harpacticoida.

Table 4. Gregarious and associative settlement

Systematic group	Genus or species	Reference	Evidence	Function	Notes on settlement inducer
Protozoa	*Folliculina*	Andrews, 1923, 1949	S	G	Adhesive trails
Parazoa	*Lissodendoryx* *Ophlitaspongia*, etc.	Delage, 1892 Wilson, H. V., 1907 Burton, 1949 Fry, 1971	S, D	G	Coalescence during or after settlement
Hydrozoa	*Aglaophenia*	Prenant and Teissier, 1924	S	H	On *Halidrys* and *Cystoceira*
	Clava squamata	Williams, 1965	S, D	H	On axils of *Ascophyllum*
Gastrotricha	*Turbanella hyalina*	Gray and Johnson, 1970 Boaden and Erwin, 1971	S, D, B	H	Deposit microflora. Avoids substance produced by *Protodrilus symbioticus*
Anthozoa	*Alcyonium coralloides*	Bourdillon, 1954	S, D, Ex	H	On gorgonian skeleton and *Mytilus* periostracum
Archiannelida	*Protodrilus rubropharyngeus* (adults and larvae)	Jagersten, 1940 Gray, 1967b	S, D, B	H	Shell gravel microflora
	P. symbioticus *P. hypoleucus* }	Gray, 1966a, b, c; 1967a	S, D, B	H	Deposit microflora
Polychaeta	*Scolecolepis fuliginosa*	Day and Wilson, 1934; Gray, 1971	S, D, B	G, H	Bacterial film and exudate of own species
	Notomastus latericeus	Wilson, 1937	S, D	H	
	Ophelia bicornis	Wilson, 1948, 1952, 1955	S, D, B	H	Sand bacteria
	Sabellaria sp.	Wilson, 1968, 1970a, b	S, D	G	Contact with tube building cement
	Spirorbis spirorbis (= *S. borealis*)	Knight-Jones, 1951 De Silva, 1962 Knight-Jones *et al.*, 1971 Williams, 1964	S, D, Ex	G, H	*Fucus* and related genera
	S. pagenstecheri	Knight-Jones, 1951	S, D	G	On stones
	S. corallinae	De Silva, 1962	S, D	H	*Corallina officianalis*
	S. rupestris	Gee and Knight-Jones, 1962; Gee, 1965	S, D, Ex	H	*Lithothamnion polymorphum*

Polyzoa	*Membranipora membranacea*	Ryland, 1962b	S	H	Blade of *Laminaria* (flexible zooaria)
	Electra posidoniae	Gautier, 1954	S	H	*Posidonia oceanica*
	Scrupocellaria reptans *Crisia eburnea*	Stebbing, 1971	S	H	*Flustra foleacea* Proximal (older) part
	Scrupocellaria scruposa *Crisia aculeata*	Stebbing, 1971	S	H	*F. foliacea* Distal (younger) part
	Valkeria uva, *Amathia lendigera,* *Mimosella gracilis*	Prenant and Teissier, 1924; Ryland, 1962b	S	H	*Halidrys siliquosa*
	Alcyonidium polyoum, *Alcyonidium hirsutum*	Ryland, 1959; Crisp and Williams, 1960	S, D, Ex	H	Fronds of *Fucus* sp. esp. *F. serratus*
	Watersipora cucullata	Wisely, 1958a	S, D	G	
Bivalves					
	Teredo sp.	Harington, 1921; Isham and Tierney, 1953	S, D	H	Wood
	Crassostrea virginica	Crisp, 1967; Hidu, 1969; Veitch and Hidu, 1971	S, D, Ex	G	Body extract, shell liquor
	Ostrea edulis	Cole and Knight-Jones, 1949; Knight-Jones, 1952; Bayne, 1969	S, D, Ex	G	Muscle extract; extrapallial fluid
Gastropoda	*Nassarius obsoletus*	Scheltema, 1961	S, D, Ex	H	Mud extract
	Adalaria proxima	Thompson, 1958	S, D	F	*Electra pilosa*
	Tritonia hombergi	Thompson, 1962	S, D	F	*Alcyonium digitatum*
	Trinchasia aurantia	Swennen, 1961	S, D	F	*Tubularia indivisa*
	Eubranchus exiguus	Tardy, 1962	S, D	F	*Kirchenpauaria pinnata echinulata*
Crustacea Copepoda (interstitial harpactacoid)	*Leptostacus constrictus*	Gray, 1968	S, D, B	H	Sand microflora

Table 4 (*contd.*)

Systematic group	Genus or species	Reference	Evidence	Function	Notes on settlement inducer
Cirripedia	*Balanus balanoides* *Balanus crenatus* *Elminius modestus*, etc.	Knight-Jones and Stevenson, 1950; Knight-Jones and Crisp, 1953; Knight-Jones, 1953b, 1955; Crisp and Meadows, 1962, 1963; Gabbott and Larman, 1971	S, D, Ex	G	Integumental proteins "arthropodin"
	Pyrgoma anglicum	Moyse, 1971	S	H, G	*Caryophyllia smithii*
	Chelonibia patula	Crisp (unpublished)	S	H, G	*Callinectes sapidus*

Key 1. Evidence:

- S. Settlement pattern: (a) animals clustered, or spat settled predominantly among its own adults or (b) animals found in association with or settling on host species.
- D. Experimental demonstration of settlement induced by its own or another species.
- Ex. Extract shown to be effective and contain settlement factor.
- B. Bacterial inoculum shown to transfer settlement factor.

2. Primary function:

- G. Gregarious settlement.
- H. Habitat recognition through association.
- F. Food recognition by settlement on prey.

These responses to the chemical nature of the substratum or to part of it are not only widespread but are also by far the most specific and usually the most powerful releasers of settlement behaviour and metamorphosis. Sad to say from the point of view of a solution to the antifouling problem, substances that have the opposite effect, namely of delaying or inhibiting settlement, are either very rare or have escaped notice. They may well be unnecessary since the majority of larvae will not settle readily in the absence of a releaser. One can only point to the fact that the surfaces of certain organisms appear to be remarkably free from external growths when alive, while others are not. The majority of these naturally protected surfaces depend on mechanical devices which disturb or inconvenience would be epizoites. The means used are characteristic of certain phyla; the echinoderms have minute snappers (pedicellariae), the Bryozoa have beaks and whip-like organs (avicularia and vibraculae) and the Hydrozoa mobile stinging organs (dactylozooids and nematophores). Other defences include mucus secretion, which is particularly widespread in the animal kingdom and occurs on many algal surfaces, while the arthropods' habit of sloughing the integument at ecdysis may also have an antifouling function (Crisp and Patel, 1958). Of chemical defences, only one example comes to mind: Sieburth mentions that bacteria are discouraged from attaching to certain plants by secretions of phenols and acid materials. However, interspecific aggressiveness between the hydroids *Obelia dichotoma* and *Clytia volubilis* cohabiting *Sargassum* weed is probably due to chemical recognition (Kato *et al.*, 1961, 1963, 1967), while similar intraspecific aggression between established individuals is shown by genetically dissimilar anemones.

VII. Nature of Chemical Inducements

Chemical stimulation of larval settlement has been demonstrated in cirripedes, bivalves, bryozoans and polychaetes. Investigations of gregariousness in the barnacle *Balanus balanoides* will be described in some detail first, since the isolation and characterization of the active principle has been carried furthest in this species.

A. Barnacles

Moore (1939) noticed how slowly barnacles colonize newly exposed surfaces and Burton (1949) gave reasons for believing that *Balanus balanoides*, among a number of other marine organisms, would be found to settle gregariously. It was also well known that the hermaphrodite individuals of species of the stalked barnacle, *Scalpellum*, which live in a more or less isolated condition in deep water, are nearly always associated with dwarf complemental males. The latter must therefore be capable of seeking out and recognizing the former. However, it was not until 1950 that gregarious settlement in barnacles was first clearly demonstrated. Knight-Jones and Stevenson (1950) showed that *Elminius modestus* settled on collectors exposed in the sea in much greater numbers if the spat was already present on them. Knight-Jones (1953b) examined the phenomenon in detail under laboratory conditions, taking advantage of the abundance and dominance of wild cyprids of *Balanus balanoides* which occur in the Menai Strait plankton from mid-April to mid-May. He showed that this species settled readily on surfaces bearing its own adults, and scarcely at all on virgin surfaces. By testing various fragments of barnacle tissue, he found that to initiate settlement it was necessary for the cyprid to make physical contact with the surface of the cuticle, shell, or base of a living or dead barnacle. Moreover, after doing so, there was a change in the behaviour of the cyprid (Crisp, 1961).

Residual bases of detached individuals provided convenient subjects for chemical treatment. Knight-Jones showed that they would stand treatments as severe as being heated to 200°C, refluxed with powerful organic solvents, immersed for a few minutes in cold concentrated sulphuric acid or strong alkali, and attacked by mild oxidizing agents (hydrogen peroxide), proteolytic enzymes (pepsin) and coagulators (formaldehyde) without losing all their efficacy. Only agents so highly destructive that they would be likely to remove the bases altogether were found to eliminate the stimulus. These included temperatures above 300°C, hot concentrated mineral acids, boiling caustic alkalis and sodium hypochlorite. Moreover, the bases of barnacles which had been preserved in formalin for six years and given prolonged washing in sea water were highly active in stimulating settlement. These properties, together with the evidence of the importance of contact with the cuticle, led to the conclusion that the origin of the stimulus must lie in the recognition of the specific character of the tanned protein of the epicuticle. Chemosensory perception of a diffusible water-soluble substance emanating from the cuticle seemed most unlikely.

Despite this conclusion, we attempted to extract a settlement inducing substance. To our astonishment, we were able to obtain crude aqueous extracts, not only from barnacle tissue but also from the tissues of all other arthropods tested as well as from one or two quite unrelated forms, which were highly active when applied to slate panels (Crisp and Meadows, 1962). The panels so treated did not lose their activity after having been washed for many hours in water, suggesting that the water-soluble material had become strongly adsorbed on to the surface of the slate. When subjected to most of the destructive agents which real barnacle bases could withstand, such slates retained their activity almost to the same degree. It seemed probable, therefore, that the material extracted from barnacle tissue was identical with, or very similar to, that which occurs in the barnacle cuticle and to which the cyprid normally responds. A major advance was now possible as a result of the technique of transferring material by extraction and adsorption; the settling factor could be subjected to chemical processes and the resulting solutions could be assayed for activity by immersing standard pieces of slate in the solution or by applying the solution to the slate surface. Ten small hemispherical pits were always drilled in the slates to indulge the cyprids' liking for concavities. A further advantage was only a few drops of solution were needed for bioassay. It was necessary to coat only the surface of the pits, 3–4 sq. mm each, and not the whole surface of the panel; cyprids will settle very readily in such pits (Fig. 7). It was shown that the settling factor was stable to boiling, was not dialysable, and was unaffected by formaldehyde. It could be precipitated by strong ammonium sulphate and the amount of dissolved protein measured chemically always kept in step with the activity assayed biologically.

There exists in the integument of all arthropods a group of proteins called "arthropodins". They can be extracted readily from soft, freshly formed integuments. In most arthropods, polyphenols and polyphenol oxidases produce quinones which harden the new cuticle soon after it has been laid down by cross-linking the molecules of arthropodin; the cuticle does not then yield as much soluble protein on extraction.

We adopted the theory put forward first by Knight-Jones that the tanned insoluble form of the arthropodin molecule could be sensed by the exploring cyprid and was the natural settlement factor. When the cuticle is homogenized in water, the arthropodin which is still untanned can leach out and this constitutes the soluble settlement factor. When physically adsorbed on to a slate surface, it presents to the water a configuration just like the cuticular surface of the barnacle and this deceives the cyprid into giving the normal gregarious response. Some further evidence was provided by physical tests which confirmed that the settlement-inducing protein met the solubility criteria that Trim (1941) had given for insect arthropodin. But much more convincing were the results obtained from the biological approach. First, it was shown that the settlement factor derived from the crab *Carcinus* was present in its most concentrated form in the integument. Secondly, considerably more settlement factor per gram of tissue could be obtained

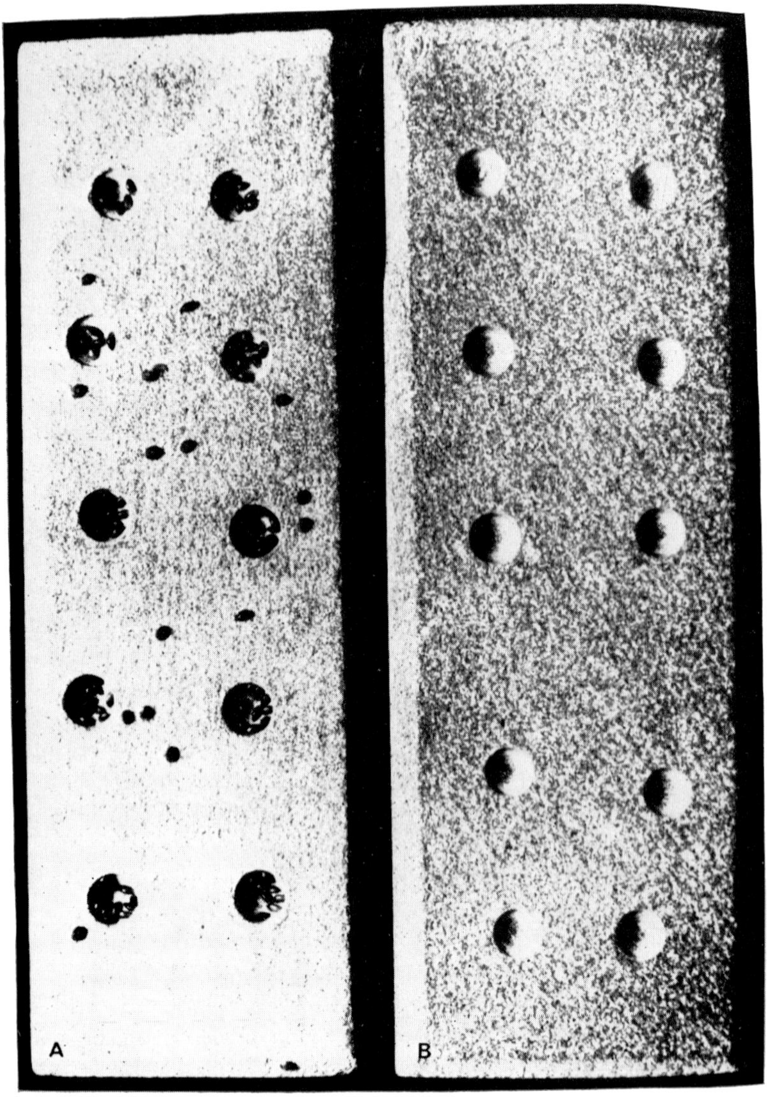

Fig. 7 Influence of barnacle settlement factor on cyprids of *Balanus balanoides*. A: slate panel cleaned with hypochlorite and soaked in a distilled water and a solution of partially purified settlement factor. B: identical type of panel treated as control, similarly cleaned but without settlement factor in the soaking bath. Both panels were offered to cyprids in the same dish. (From Crisp and Meadows, 1963.) (Reprinted with permission of the Royal Society.)

from recently moulted animals than from those whose cuticles had hardened. Thirdly, arthropod tissues with soft integuments, such as the stalks of *Lepadidae* or recently settled "pinhead" barnacles were very rich in settlement factor.

The provenance of substances which show similar settlement inducing activity from the integument of other arthropods is less surprising than the presence of settlement inducing activity in extracts of some sponges and of the fish *Blennius pholis*. At first, contamination was suspected, since barnacles often settle among patches of the sponge *Halichondria panicea* and *Blennius* feeds on barnacles. More careful experiments, in which all arthropod tissue was carefully excluded, confirmed that settling substances did exist in the extracts of these animals. These settling substances would probably be quite different from those derived from arthropods and in fact the crude extract derived from the sponge lost appreciable activity after five minutes boiling (Table 5),

Table 5. Stability of *Halichondria* extract to boiling

Aqueous extract from 3 parts sponge by vol. to 1 part sea water

Panel treatment	Mean settlement and standard error from 6 replicates
Control (clean)	0·3 ± 0·2
Balanus balanoides standard crude extract	32·3 ± 2·0
Halichondria extract × 1	15·3 ± 2·7
Halichondria extract × 1/10	13·5 ± 1·5
Boiled *Halichondria* extract × 1	5·3 ± 1·6
Boiled *Halichondria* extract × 1/10	2·7 ± 1·1

whereas a corresponding barnacle extract was unaffected over this period of time (Table 6).

There may be an evolutionary significance to some of the anomalously active extracts (Crisp, 1965). It is well known that although the majority of species settle on rocks and other inert substrata, certain groups of barnacles have evolved an epizoic habit, normally settling on the surfaces of other groups of organisms. The commonest living substrata to be thus infested are the surfaces of corals and Gorgonia by *Pyrgoma*, *Creusia* and *Conopea*; the tissues of sponges by some species of *Balanus*, and *Acasta*; the surfaces of aquatic vertebrates—whales, turtles, sea snakes, manatees, sharks, etc.—by *Coronula*, *Platylepas*, *Tubicinella*, *Xenobalanus* and *Anlasma*, and of course other crustaceans, including barnacles, by *Chelonibia*, *Conchoderma*, *Poecilasma*, *Octolasmis* and *Ibla*. Substances inducing settlement of *Balanus balanoides* cyprids have been found in representatives of all these groups except the coelenterates which have not been exhaustively tested. So far, negative results have been obtained with *Caryophyllia smithii*, the host of *Pyrgoma anglicum*, and a gorgonian to which *Conopea galeatus*

Table 6. Effect of boiling partially purified arthropodin in water and in normal hydrochloric acid

Barnacles extracted with minimum quantity of distilled water, crushed and boiled, centrifuged to 5000 r.p.m. for 30 min to remove solids. Protein precipitated from supernatant by addition of sat. ammonium sulphate, centrifuged, and dialysed 12h in running tap water, followed by 24h in changes of distilled water. Solutions withdrawn at intervals during boiling and tested by soaking six replicate slate panels in each test solution and offering to cyprids.

	Boiled in Water		Boiled in N. HCl		
Time	Number settled on panels soaked in resulting solution	Percentage of settlement on original	Numbers settled on panels soaked in resulting solution	Percentage of settlement on original	Percentage of spreadable protein left after acid hydrolysis
0	160	100%	83	100%	100%
5 min	—	—	50	60%	92%
30 min	130	81%	12	14%	59%
2h	161	100%	0	0%	28%
6h	126*	79%	1	1%	26%
20h	94**	59%	0	0%	15%
Controls soaked in sea water	1	(0·6%)	0	0%	

* Slight precipitate formed; ** Considerable brown precipitate formed

is often attached. Since sponges and vertebrates possess substances which can present molecular arrangements sufficiently similar to those of the arthropod cuticle to stimulate cyprids of non-epizoic species, it seems quite likely that, where biochemically favourable surfaces exist on an organism that would provide an ecologically suitable habitat, an epizoic and perhaps ultimately a parasitic relationship may well have evolved.

Some hitherto unpublished investigations by Dr. D. A. Ritz and Dr. P. A. Gabbott have confirmed the protein constitution of the settling factor and emphasized the importance of its integrity. It is not affected by lipid solvents, nor by carbohydrases, but hydrolysis with boiling 1·0 normal hydrochloric acid resulted in loss of activity as judged by the numbers of cyprids settling on panels soaked in the resulting partial hydrolysate. The loss took place quite rapidly, being virtually complete within the first 30 minutes (Table 6) and it could not have been caused by heat denaturation since boiling in water has no corresponding effect. It is important to note that the numbers of cyprids settling on test panels are not necessarily proportional to the concentration of the active principle present in the solution used to treat the panels. As the experiment on *Halichondria* extract shows (Table 5), dilution of an extract has only a relatively small influence on the numbers settling. Such would be expected if the surface were nearly saturated with adsorbate. Therefore, the reduction in activity caused by boiling hydrochloric acid probably implies a much more rapid loss of active material than the settlement figures indicate. The breakdown of the macromolecules into smaller units during hydrolysis could be followed by the Langmuir trough technique, since the original ammonium sulphate purified extract spreads as a surface film on water (Crisp, 1965). The rate of loss of spreadable material is shown in Table 6. It is clearly less rapidly lost than the active principle. Tryptic and peptic digestion was similarly investigated but the monolayer technique was not appropriate because

Table 7. Hydrolysis of barnacle arthropodin by enzymes. Preparation of material as described in Table 6.

(1) Tryptic digest 37°C, 0·25M Phosphate buffer pH = 7·4

Time	Settlement as % of that on original	Amino nitrogen not released from protein (by Pope and Stevens method)
0	100%	100%
6 min	15%	93%
15 min	1%	90%
180 min	0·5%	85%

(2) Peptic digest at 37°C, 0·2M Acetate buffer pH = 2·0

Time	Settlement as % of that on original	Amino nitrogen not released from protein (by Pope and Stevens method)
0 min	100%	100%
6 min	100%	93%
10 min	73%	91%
60 min	69%	90%

of possible interference from the enzymes added. Instead, the breakdown of the macromolecule was followed from the increase in α amino nitrogen, which was measured using Pope and Stevens' method. I am indebted to Dr. A. O. Christie for carrying out the analysis. The findings

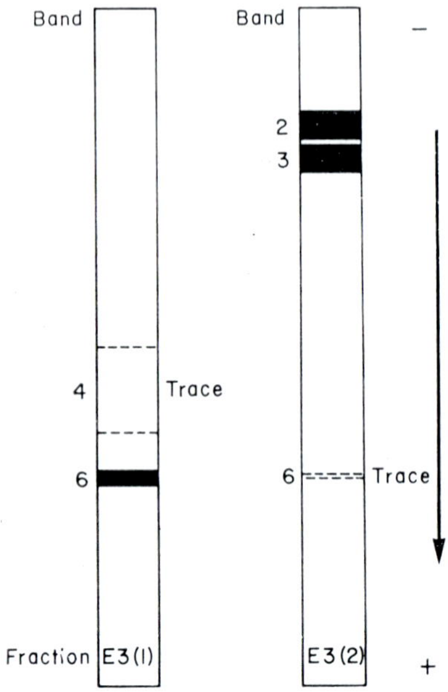

Fig. 8 Diagrammatic representation of electrophoretic patterns of a partially purified barnacle settlement factor (E3), separated by preparative electrophoresis into faster and slower moving fractions E3(1) and E3(2). Electrophoretic conditions pH 9·2, 5% acrylamide gel, 6mA/tube for 45 minutes. Three main bands, a fast moving band 6, the main component of E3(1), and two slowly moving bands 2 and 3 the main components of E3(2). Arrow, direction of migration. (From Gabbott and Larman, 1971.) (Reprinted with permission of Cambridge University Press.)

in Table 7 again demonstrate that the loss of activity was rapid compared with the rate of breakage of the protein chain.

These results imply that if the arthropodin molecule is broken into smaller units, though most of them are still macromolecular and able to be spread on water, they have completely lost their activity. The molecule needs to remain intact in order to be recognized; either the removal of small essential pieces or its disintegration into a few subunits destroys its activity. The activity is therefore unlikely to be caused by particular side chains or by frequently repeated small sections of the molecule. Bioassays carried out by Mr. P. S. Meadows with synthetic polymers of alanine, tyrosine, lysine and glutamic acid were all negative, which tends to give support to the conclusion that sequences of these side-chains are not involved.

Gabbott and Larman (1971) refined the fractionation procedure by batchwise treatment of the resuspended ammonium sulphate precipitate with DEAE cellulose to remove nucleic acid and then isolated the molecular weight fraction in excess of 50,000 with sephadex G50. Preparative gel electrophoresis divided the material into two fractions, E3(1) and E3(2) which revealed on acrylamide gel three main bands. The fast band (6, Fig. 8) was the major component of the first fraction, E3(1), and stained for protein only. The two slower bands (2 and 3, Fig. 8) which constituted most of the second fraction, E3(2), stained positively with amido black 10B and P.A.S., hence contained protein and carbohydrate. When the two fractions were tested against cyprid larvae, both induced settlement and it followed that at least two active components were present in the original extract. One was probably a glycoprotein or mucopolysaccharide-protein complex and the other a protein which had been associated originally with nucleic acid. More recently, V. N. Larman has shown that the fast moving component (band 6, Fig. 8) is polydisperse and separable into two groups of subfractions, a prominent group with isoelectric points in the range 4·5–5·5, and a less prominent group with isoelectric points above this range but below pH 7·0, as shown in Fig. 9. Extracts from different barnacles, though differentiated by the cyprids, yield, after purification and separation by isoelectric focusing, patterns identical in appearance with each other. Extracts from the crab *Carcinus* also give similar patterns.

The technique of transferring the settlement factors from tissues to test surfaces, which was used first with barnacles, has now been applied to many other sessile organisms. However, unless the effect is both profound and specific it may sometimes be difficult to discriminate between the influence of the settling factor and that of bacterial films which build up rapidly when a surface is immersed in a non-sterile organic rich medium. The difficulty is minimized when dealing with larvae which are repelled by or are uninfluenced by bacterial films. This appears to be the case with some polyzoan larvae.

B. Polyzoa (Bryozoa or Ectoprocta)

Capitalizing on the evidence obtained by Ryland that the Bryozoa *Alcyonidium* and *Flustrellidra* selected the fronds of *Fucus serratus*, Crisp and Williams (1960) prepared crude extracts of this and other algae in which experimental panels were steeped. Essentially, three types of surface were prepared and offered simultaneously to the polyzoan larvae: (a) a clean surface, (b) a surface soaked in sea water for 24

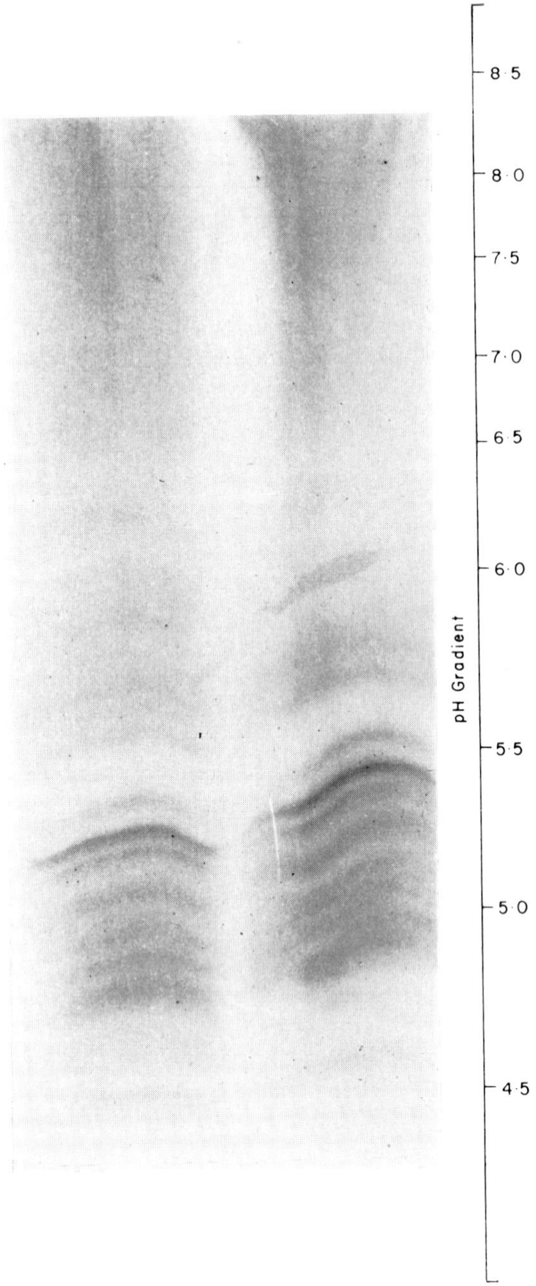

Fig. 9 Separation of fast moving fractions of two barnacle settling factors by isoelectric focusing. Flat-bed acrylamide gel isoelectric focusing applied to partially purified extracts of barnacle settlement factor prepared as Gabbott and Larman (1971). A. *Balanus balanoides*, B. *Elminius modestus*. Samples were applied near cathode after pre-focusing the plate for 2 hours. Plate was run at constant current limiting voltage 400V. Current/voltage at start of pre-run 10 ma 215V dropping to 4 ma 400V after 2 hours. The carbon electrodes were 18 cm apart. Anode was moistened with 5% H_3PO_4 and the cathode with 5% ethanolamine. The gel was polymerized with riboflavine and contained 1·6% ampholytes (LKB Ltd) pH range 3–10. V. N. Larman, unpublished. Photograph: D. C. Williams.

hours to acquire a bacterial film and (c) a filmed surface soaked in *Fucus serratus* extract. Table 8 shows that the larvae of *Alcyonidium polyoum*

Table 8. Response of larvae to panels treated with extracts of *Fucus serratus*. (From Crisp and Williams, 1960, and Williams, 1964)

Treatment of surface	Settlement of larvae of *Alcyonidium polyoum*	*Spirorbis spirorbis*
	(Series 1)	(Series 2)
(a) Clean unfilmed	11	74
(b) Filmed in sea water	5	307
(c) Filmed and soaked in algal extract	162	6475
Total settlement on all panels in each series of experiments	178	6856

did not settle to any great extent on surfaces (a) or (b) but surface (c) was heavily colonized. Extracts of other fucoid algae were found to encourage settlement but algae unrelated to *Fucus serratus* did not.

It should also be seen from Table 8 that the larvae settled in slightly smaller numbers on the filmed surface than on the clean surface. It is unlikely, therefore, that the additional attractiveness produced by the extract results from a mere increase in the thickness of an organic film; it must be ascribed to a qualitative difference. Incidentally, the presence of a bacterial film enhanced the influence of the extract, perhaps by acting as a reservoir or absorbent for the active principle. Otherwise, little is known of the mechanism or the nature of the active principle. It is certainly not the mucus which these brown algae produce so freely.

C. Polychaetes

Table 8 also summarizes the results of a number of identical experiments performed by Williams (1964) using the larvae of *Spirorbis spirorbis*. This larva prefers filmed surfaces and therefore the assumption that the increase in the amount of film-forming material can be discounted as a factor in settlement no longer applies. It was necessary, therefore, for Williams to measure the film forming material (as organic matter) present on glass slides treated by each of these three methods. To improve the sensitivity, he measured the organic matter on the surface of powdered glass which had been filmed in sea water, filmed and treated with *Fucus* extract, and treated with a peptone-glucose enrichment culture in the same way as glass slides used for settlement experiments. He found that the organic carbon was increased only from 138–207μg/g (μg Carbon/g. glass) after immersion in *Fucus* extract, whereas the glass kept in enrichment medium acquired 718μg/g of carbon. The slides treated with *Fucus* extract were still overwhelmingly preferred to all the others, in fact the surface heavily coated with a bacterial film was not settled on at all. Within this series of experiments, treatment with *Fucus* extract rendered the surfaces attractive to *Spirorbis* without prior filming in sea water. Moreover, experiments done under virtually sterile conditions in pasteurized sea water showed that panels treated with boiled or membrane filtered *Fucus* extract gave a strong stimulus to settlement. There seems little doubt, therefore, that the associative settlement of *Spirorbis spirorbis* on *Fucus* is due to a chemical response to a material naturally present in the *Fucus* fronds.

The settlement of *Spirorbis rupestris* in the vicinity of the red calcareous alga, *Lithothamnion*, can be simulated by soaking slate panels in an aqueous extract of the alga (Gee, 1965). The settlement factor would withstand being dried but was relatively unstable to boiling and to formaldehyde treatment. It appeared to pass freely through a dialysis membrane and therefore to be of fairly small molecular weight.

Although the study of associative settlement has now been fairly well advanced, nothing is known of the factors responsible for gregariousness which has been demonstrated in at least two species of *Spirorbis*, *S. spirorbis* and *S. pagenstecheri*. The former shows both gregarious and associative settlement and it would be interesting to know whether the substances involved have any chemical or physical similarity.

Though Wilson's investigations (1968, 1970a, b) on the settlement inducing substance of the polychaete *Sabellaria* have not yet reached the stage of chemical analysis, the analogy with the barnacle settling substance is very close. Wilson looked at two species, *Sabellaria spinulosa* and *S. alveolata*. The former is a northern species which forms sublittoral colonies in the North Sea but in the south-west of Britain lives in isolated tubes. *Sabellaria alveolata*, which is restricted to the milder west coasts of Britain, always lives gregariously, building large if somewhat fragile reefs attached to rocks in the lower intertidal zone of exposed sandy shores. Both produce planktonic larvae with long pelagic phases. In laboratory cultures of *S. spinulosa* and *S. alveolata*, the normal time to reach the settling stage is 2–3 and 2–4½ months respectively, while settlement can be postponed up to at least 7 and 9 months. Clearly, the larvae would be very widely disseminated in the sea during this period so that a gregarious response at settlement is essential. Wilson found that both species could detect the cement which is secreted on to the grains to bind together the particles used in tube building. This material can be recognized only on contact, it is insoluble in water, resistant to drying but can be destroyed by cold concentrated mineral acids. According to Vovelle (1965) it is a proteinaceous material formed from two types of gland cells and is cross-bonded by quinone tanning. *S. spinulosa* larvae show great specificity, distinguishing clearly between their own tube fragments and those of *S. alveolata* (Wilson, 1970b) but the larvae of *S. alveolata* behaved paradoxically. When offered pieces of otherwise identical sand, one pile of which had been used to build tubes in the laboratory by *spinulosa* and the other by *alveolata*, in four out of five experiments the larvae of *S. alveolata* unexpectedly chose *spinulosa* sand (Wilson, 1968). However, when offered a choice between sands from naturally occurring tubes they settled more often into piles of *alveolata* sand. The ranking order of sands with *Sabellaria* cement appeared to be (1) used by *S. spinulosa* to build tubes in the laboratory, (2) used by *S. alveolata* to build tubes in the laboratory (3) used by *S. alveolata* to build tubes in the sea, (4) used by *S. spinulosa* to build tubes in the sea. Quite probably, the freshly secreted cement is a more powerful stimulus than cement aged in the sea.

D. Oysters

With the development of hatchery methods for commercially important bivalves and the need to produce artificial seed, methods of promoting settlement on collectors are likely to be valuable. Cole and Knight-Jones (1949) had already shown experimentally that larvae of *Ostrea edulis* set gregariously and Knight-Jones (1952) demonstrated that proximity to oyster layings enhanced the settlement on cultch in the sea. Attempting to show an analogy with barnacles, I stripped the periostracum layer from *Crassostrea* shells with sodium hypochlorite and compared their performance against that of shells retaining the periostracum when offered to the larvae of *Crassostrea virginica*. Fortunately, oyster larvae are not very sensitive to surface texture so that possible changes in the roughness of the shells brought about by chemical treatment appeared to be insignificant. There was a clear preference for shells which retained the natural tanned protein layer (Crisp, 1967). Closely similar results have been obtained by Dr. D. A. Ritz with larvae and matched shells of *Ostrea edulis*, and these are displayed, together with the experimental details and statistical analyses, in Table 9. Larvae of *Ostrea edulis* therefore behave similarly to those of *Crassostrea virginica*.

Extracts made from oyster tissue also promote the settlement of larvae but, surprisingly, the effect was obtained not only on shells stripped of their natural protein layer but equally on natural shells

Table 9. Effect of removal of periostracum of adult shell to settlement of larvae of *Ostrea edulis*

Pairs of matched shells, one treated with hypochlorite for 24h, the other with anionic detergent. Both shells well washed and scrubbed. Half the pairs (24) dried and half soaked 24h in sea water to remove any trace of hypochlorite and coat with surface film before being offered to larvae. Numbers counted on inner and outer surfaces of all 48 shells. Analysis of variance of results after log transformation to homogenize variations. Number of spat on stripped shells, 914; on natural shells, 2750.

Factor tested	No. of values	Log_{10} mean no. cm^{-2}	Mean no. cm^{-2}
(1) Removal of periostracum:			
all shells without periostracum	48	$\bar{1}$·34 *	0·22 *
all shells retaining periostracum.	48	$\bar{1}$·85 *	0·71 *
(2) Bacterial film:			
all shells scrubbed and dried;	48	$\bar{1}$·54	0·35
all shells soaked in sea water 24 hours.	48	$\bar{1}$·66	0·45
(3) Inside and outside of shell:			
all inner surfaces;	48	$\bar{1}$·54	0·35
all outer surfaces.	48	$\bar{1}$·66	0·45
Grand mean	96	$\bar{1}$·60	0·40

* significantly different

Analysis of variance;

Source of variation:	Degrees of freedom	F ratio
with and without periostracum;	1	52·2
Residual.	94	1

Significant difference at P <0·05 for each of the above means of 48 observations = 0·14

which were already partially covered with periostracum (Crisp, 1967). Unfortunately, the effect of a chemical inducement is much more difficult to demonstrate in oysters than in barnacles because the settlement of the pediveliger gives a ratio of only about 3:1 between treated and untreated surfaces instead of 100 or even 1000:1 achieved by the more discriminating cyprid. In view of the great variability experienced in all settlement experiments, results in which a ratio of less than 2:1 is obtained are not usually reliable and this sets a difficult target for critical experiments with oyster larvae. Notwithstanding, some very interesting observations on the nature of the settling factor have been made by Bayne on *Ostrea edulis* and by Hidu and his collaborators on *Crassostrea virginica*. Bayne (1969) obtained from the adductor muscle and the extra-pallial fluid an extract containing high molecular weight material stable to boiling but destroyed at a slightly acid pH. It could be purified by precipitation with ammonium sulphate, but the purified material was less stable, coagulating on boiling, acidifying or freezing though it could be dried or precipitated by trichloracetic acid without significant loss. The material could be destroyed by pronase and contained mainly protein, but hexose sugar elements were also present in significant amounts with some lipid and traces of hexosamine, uronic and nucleic acids, probably as impurities. Bayne found that *O. edulis* larvae were specific in their response, being stimulated only to a limited extent by extracts of *Ostrea lutaria* and not at all by those of *Crassostrea* and *Mytilus*. Veitch and Hidu (1971) obtained and similarly purified from the oyster shell liquor of *Crassostrea virginica* a settling factor which also gave a protein and a faint carbohydrate reaction. They found that only fractions of molecular weight of 100,000 or more were active. They also claimed that solutions containing iodinated amino acids and proteins encouraged settlement, but the ratios between the numbers settled on experimental and control surfaces did not exceed 2·0. Moreover, great caution is needed in the interpretation of results carried out in different vessels with different batches of larvae. Hidu (1969) thought that a soluble factor was present, but Bayne (1969) maintained that a contact stimulus was necessary; this question will be discussed below.

VIII. Mechanism of Larval Response to Chemical Stimuli

It would at first sight seem probable that chemical recognition would imply some form of olfaction, the larva being able to respond to small quantities of attractant diffusing from the surface. Two examples in which this appears to take place have been described earlier (Scheltema, 1961; Harington, 1921) and there are several others in the literature. Bourdillon (1954) showed that by putting an extract of a gorgonian into the water the settlement of *Alcyonium coralloides* increased compared with that in sea water alone or even in the presence of the gorgonian itself. Thompson (1958) caused the larvae of *Adalaria proxima* to metamorphose in the presence of live colonies of *Electra pilosa*, the adult prey and the normal settlement releaser, even though they were separated by plankton silk which would have prevented direct contact. Hidu (1969) used spat of *Crassostrea* contained in bags to stimulate settlement of *Crassostrea* larvae and concluded that a water-borne pheromone must be produced. It is just possible that in oysters there are two separate phenomena, one by which th elarva recognizes the protein of the shell surface, and another by which it responds to soluble material dissolved from the tissues of the adults. This would account for the fact that settlement on natural shells on which the periostracum is

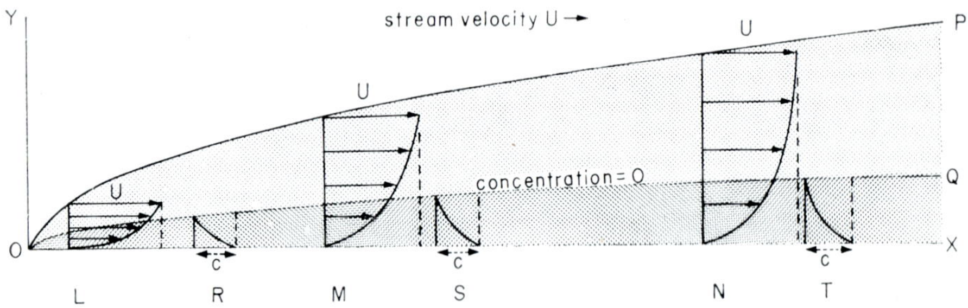

Fig. 10 Diagram of the diffusional and velocity boundary layers formed when water flows past a flat plate. A substance is diffusing from a constant concentration at the plate surface to virtually zero concentration at the outer boundary of the heavily shaded area. The concentration profiles are shown within it. The hydrodynamic boundary layer is shown by lighter shading and velocity profiles from zero to the stream velocity U outside the plate are shown. The hydrodynamic boundary layer lies outside the concentration boundary layer provided that the diffusion constant is less than the kinetic viscosity, as is normally true for aqueous media. For small objects of 1–10 cm and fluid velocities of 10 cm/sec the diffusion boundary layer is less than a millimetre thick, comparable with the dimensions of an invertebrate larva. (From Crisp and Meadows, 1962.) (Reprinted with permission of the Royal Society.)

intact can be further improved by soaking in oyster extract (Crisp, 1967) as well as the observations of Knight-Jones (1952) that a greater natural spatfall was obtained in the vicinity of living oysters and spat than elsewhere, even on the surface of single valves that have long been empty.

A difficulty which was first stressed by Harington and later put forwards in quantitative form by Crisp and Meadows (1962) is that a soluble pheromone would be diluted to extinction within a very short distance of the surface to which the larvae were being attracted (see Fig. 10). The "mud factor" described by Scheltema (1961) which causes *Nassarius obsoletus* to metamorphose, would be free from this objection since it is quite conceivable that enough material might diffuse from the large expanses of bottom deposit present in the estuary significantly to influence the chemical composition of the water and thereby cause the larvae to modify their swimming behaviour and settle out. But the possibility of a diffusant from a small area of surface being able to attract larvae from a distance seems extremely remote except in an artificially contrived experiment where the effects of dilution through convection are minimized. Indeed, if such a mechanism were effective in the sea, it should be very easy to reproduce it in small vessels in the laboratory where the concentration of the pheromone would be much higher. Yet, where larvae have been seen to settle in response to a chemical stimulus, most observers have remarked on the absence of any indication that the larva is influenced from a distance. Thus Bayne (1969) could detect no change in behaviour of eyed oyster larvae until they reached a treated surface. Crisp and Meadows (1963) noticed that although the presence of the settling factor of barnacles in dilute solution altered the sign of the photic response of cyprids, it did not cause them to settle if the solution was prevented from coming into contact with the surface. Williams (1964) found that a *Fucus* extract abbreviated the pre-settlement behaviour of *Spirorbis spirorbis* but only after the larva had been in contact with a surface soaked in it. Similar observations were made on the larvae of Polyzoa. Williams suggested that in the latter case, fibres formed by the interaction between a mucus liberated by the larva and the settling factor may have promoted fixation. It is interesting at this point to note that some larvae lay mucus trails. The byssus apparatus of bivalves is essentially a post larval organ (Yonge, 1962) which normally serves to anchor the larva at settlement but which has been adapted to a number of other functions. In *Teredo pedicellata*, the apparatus produces a mucus-like material rather than a byssus for temporary attachment (Isham and Tierney, 1953). Sigerfoos (1907), probably describing this phenomenon, refers to the attachment of the larva by a single byssus thread. Larvae of *Ostrea edulis* also lay down a thread as they explore (Cranfield, 1971) and the complexity of the thread can be seen with a scanning electron microscope (Fig. 11). The larvae of at least two species of *Spirorbis*, *S. borealis* and *S. rupestris*, lay similar trails (Knight-Jones, 1951; Gee and Knight-Jones, 1962) while Williams (1964) describes how the crawling planulae of *Clava squamata* can be dislodged from the surface by hooking the mucus trail a few millimetres behind the larva. Lynch described threads associated with *Bugula* larvae. Such threads may sometimes be important in mutual recognition when other larvae make contact with the surface. They would explain why gregariousness is not necessarily related to the number of spat present (Bayne, 1969) and why a surface on which larvae have settled and from which they have subsequently

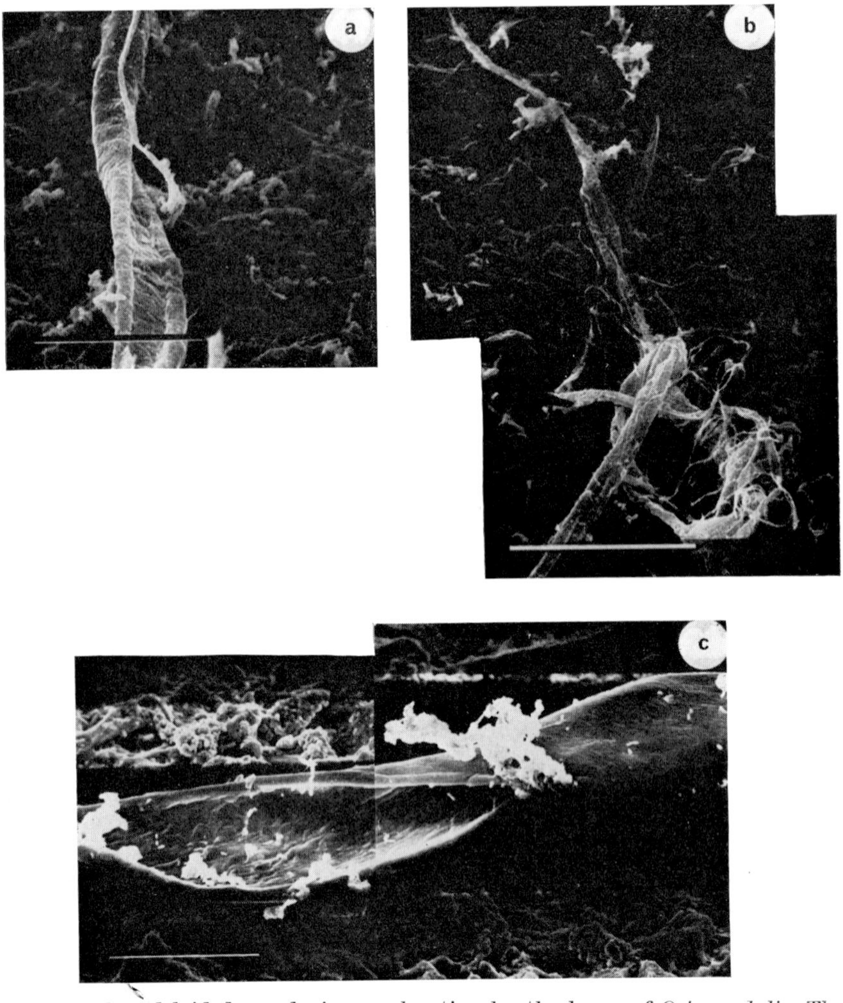

Fig. 11 Thread laid down during exploration by the larva of *Ostrea edulis*. The thread consists of two cores and an outer covering: (a) shows the bifid appearance due to the separation of the cores (Scale Bar 50μm); (b) shows the appearance when the fibrous outer layer is stripping off (Scale Bar 50μm). When the cores are widely separated the thread appears like a flat ribbon (c) (Scale Bar 25μm). Pictures by H. J. Cranfield on Cambridge Stereoscan Mark 2a Scanning Electron Microscope.

been removed is still capable of promoting settlement (Cole and Knight-Jones, 1949). Larvae of infaunal species such as *Ophelia bicornis* (Wilson, 1955) similarly need to enter the deposits and make contact with the sand grains before they are able to exercise choice and metamorphose. There is no sign of attraction at a distance. Interstitial archiannelids, which appear to have a closely similar mechanism of selection, also require physical contact (Gray, 1966a, b, 1967a, b). Species of *Protodrilus* attach to the grains by their tail lappets if the sand is favourable, but swim through the deposit when it is not.

Furthermore it is now certain that larvae can recognize such eminently insoluble organic materials as quinone tanned proteins which have been long in contact with water and are therefore unlikely to contain freely diffusible impurities. It is almost inconceivable that they could respond to molecules escaping from the cross-linked polymer itself. The matrix of such a polymer is virtually a single molecule from which it would be impossible for any part to diffuse freely away. Yet barnacle cyprids, *Sabellaria* larvae, and probably oyster pediveligers respond to such materials.

An experiment carried out on barnacle cyprids (Crisp and Meadows, 1963) confirms that the larva responds to such substances by a mechanism quite different from normal chemoreception. When the cyprids are placed in a dish with slate panels, some of which have been immersed in an extract containing the settlement factor and others which are clean, they immediately begin to attach to the former, though it may require half an hour or more before they become permanently fixed. If the panel is removed from the water after only a few minutes, they often remain attached to it; if the panel has previously been treated with settling substance, a great many remain on, whereas when clean panels are removed the majority tend to swim off. It is therefore quite possible to tell which panels are favoured without waiting for the cyprids to metamorphose. Indeed, the recognition of the settlement factor can be observed within a few minutes of placing the panels in sea water containing cyprids. Proteins are known to absorb on to a solid surface from a dilute solution rather slowly. It may require ten to fifteen minutes for an adsorbed layer of arthropodin to build up. In the experiment to be described, the arthropodin extract used was so dilute that it had no influence on cyprids swimming in it, but nevertheless panels soaked in it for twelve hours or more became very attractive to cyprids. Equal numbers of slates were used, one set having been soaked overnight in the extract and the other in sea water. Both sets were then offered to cyprids simultaneously, not in sea water but in the same dilute extract as had previously been used for treating half the panels. The cyprids unhesitatingly chose the panels previously soaked in extract.

Since all the panels were immersed in a uniform concentration of the settling factor, it was impossible for the cyprids to have distinguished between the two sets as a result of any concentration difference or diffusion gradient set up. They could have done so only by detecting the presence of an adsorbed protein layer on the surface. Moreover, they were evidently not sensitive to the settling factor in soluble form.

This result explained several earlier observations. The need for the settling factor to be adsorbed at the solid liquid interface explained why strongly polar mineral surfaces, such as slate, were found suitable for settlement experiments of this kind yet pieces of wax, polythene and P.T.F.E., etc., which lacked strong adsorbing capacity, failed to attract cyprids to any significant extent, even after treatment. It also explained why substances which tended to increase the deposition of the protein on the surface, such as ammonium sulphate and possibly formaldehyde, apparently improved the performance of an extract.

This response to adsorbed layers throws doubt on all cases where the settlement factor has been thought to act on the larva in solution, because any high molecular weight material will be adsorbed to some extent on objects in the experimental dish and on the walls of the vessel. Hence there must normally remain a possibility that the larvae could be reacting to an adsorbed layer. The only way in which one could prove that they responded to the material in solution would be by observing some change in behaviour leading to settlement or a movement along a concentration gradient of the substance, without there being any contact between the settlement factor and the surface on which the larva would eventually settle.

Crisp and Meadows termed the ability to recognize chemical structures by contact a "tactile chemical sense". The known cases all involve high molecular weight or cross-linked materials, while the majority indicate that this sense is also highly discriminating and can differentiate between very similar substances which have been made by related organisms.

An examination of analogous phenomena might be of great assistance in explaining this remarkable capacity. There is at least one other example of the recognition of tanned protein by a marine organism. Ross and Sutton (1963) noticed that the symbiotic union between the anemone *Calliactis parasitica* and the hermit crab *Eupagurus bernhardus* depended on the response of the anemone to the shell of *Buccinum* carried by the hermit. The shell factor was located both in the periostracum and in the organic matrix of the shell; the anemone seeks to plant its base on such material in preference to other substrata (Ross, 1967).

A more significant analogy with the tactile chemical sense is the antibody-antigen reaction. It resembles the gregarious response of barnacle larvae in three essentials. First, both antibody-antigen reactions and the tactile chemical sense are concerned with the recognition of large complex molecules and are sensitive to minor changes in structure. Olfaction, in contrast, is concerned with small diffusible molecules. Secondly, graded specificity with a reduction in the effect with decreasing phylogenetic affinity is characteristic of both. Thirdly, the antibody-antigen reaction is dependent on molecular forces at the outermost layers of the cell. The response of the barnacle cyprid has similarly been shown to rely on adsorption of protein to the substratum. This protein layer exerts its effect at an average thickness of between 10 and 100 Å which corresponds approximately to the thickness of a single molecular layer of protein (Crisp, 1965).

If the analogy can be sustained, it would suggest that the exploring larva is sensitive to surface forces between some part of itself and the substratum. The response might thus be mediated by forces of friction or adhesion. The alternative theory to explain the response to adsorbed layers of specific proteins is to assume that the larva is endowed with enzymes capable of breaking them down. It would then need to respond to the resulting products by the normal chemical senses. This idea was first put forwards by Knight-Jones (1953b) and re-examined by Nott and Foster (1969) after they had investigated in detail the structure of the antennulary organ of the cyprid. This organ is closely and repeatedly applied to the substratum as the cyprid explores. It is a structure packed with sense organs and therefore the most likely means by which the larva tests the nature of the surface. Nott (1969) and Nott and Foster (1969) demonstrated from its structure that it is not a sucker, as was once supposed. It has a non-living microvillous outer layer ideally suited both to penetrate into every unevenness of the surface and to become well keyed into the secreted cement when fixation takes place. Groups of unicellular glands open on to the surface of the antennulary organ and the curtain-like velum that surrounds it would help to isolate the underside when applied to the substratum. There are open-ended organs with sensory cilia opening on to its surface which could well be olfactory. Nott and Foster suggest that the glands may contain an enzyme which could break down the concentrated surface layer of the settlement substance so that the hydrolysate might then be recognized by the olfactory endings. It would be much more difficult for the animal to do this when it was swimming because the settlement substance would be many times less concentrated in solution than at an interface and the enzyme and its breakdown products would readily diffuse away.

I find this theory hard to accept, first because a complex mixture of products of hydrolysis would have to be formed and their composition recognized in the short time that the antennulary organ is applied at one place, that is, in approximately three seconds. Secondly, the differences between similar proteins derived from different species often amount to only a few residues in a large molecule. It would require an olfactory organ with supreme powers of differentiation to make the required distinctions. However, little is known at present as to exactly how olfaction works so that this objection must be a matter of opinion.

The theory which I prefer is based on the analogy of the antibody-antigen reaction and supposes that the surface of the antennulary organ possesses properties that cause it to adhere to the molecular structure of the adult epicuticle, or to the settlement factor when adsorbed on a surface with the same orientation. The behaviour of the cyprid supports this contention, since towards the end of the settlement behaviour pattern, and particularly just before cementing itself down, it can be seen to twist round and apparently pull on the antennules repeatedly. Moreover, cyprids that have alighted on panels treated with settlement substance are much more tenacious and difficult to wash off (Crisp and Meadows, 1963). The antennule possesses many sense organs which appear to be proprioceptive (Nott and Foster, 1969; Gibson and Nott, 1971) and which might well be used both for detecting curvatures and irregularities in the surface and for registering the strain on pulling the antennule. The small glands within the antennule might secrete a material for increasing the adhesion between the disc of the attachment organ and the surface on which it was planted. It is possible that it is this material that adheres particularly strongly to the arthropodin surface.

One undoubted difficulty which this theory presents is why the natural surface is not always being occluded by other substances such as bacterial films which are known to form on it. Possibly this may happen—in fact, Wilson's observations in which aged *Sabellaria* cement was less effective than fresh cement, suggests that it does. But cyprids, at least, can be seen to rub against the surface and flick off dirt at the final stage of inspection and may indeed be engaged in exposing the surface before applying cement (Knight-Jones and Crisp, 1953). It is possible also that very small gaps and imperfections in a bacterial film or in other layers overlying a surface would be sufficient to enable larval organs to penetrate and make direct contact with the substratum beneath. Possibly at the final close inspection of the site, the larvae are searching for such relatively clean areas.

Another question such a theory must eventually answer is whether the intermolecular forces between the larva and the attractive surface are likely to be perceived by the sense organs. Antibody and antigen unite together by very strong forces and when combined are often very difficult to dissociate without denaturation of one or both components. There are now in use immunoadsorbants to which antigens can be adsorbed and which then will attach strongly to the antibody, or vice versa. If the molecular interaction theory of larval response to surfaces is true, these systems offer the opportunity of studying situations closely analogous to the procedure of rendering a surface attractive by means of a settling substance. The next step is to attempt to measure the magnitude of the mechanical forces that can be generated in artificial systems and between the larva and the substratum.

References

Abe, N. (1937). Post larval development of the coral *Fungia actiniformis* var. *palaevensis*. *Palao trop. biol. Stn Stud.* **I** (1), 73–93.

Andrews, E. A. (1923). *Folliculina*; case making, anatomy and transformation. *J. Morph.* **39**, 207–78.

Andrews, E. A. (1949). Folliculinid life-cycle. *J. Morph.* **84**, 401–9.

Ankel, W. E. von and Eigenbrodt, H. (1950). Uber die Wuchsform von *Spongilla* in sehr flochen Raümen. *Zool. Anz.* **145**, 195–204.

Barnes, H. (1953). Orientation and aggregation in *Balanus balanus* (L) Da costa. *J. Anim. Ecol.* **22**, 227–241.

Barnes, H. (1956). Surface roughness and the settlement of *Balanus balanoides* (L.). *Arch. Soc. Zoo. bot. fenn. Vanamo* **10**, 2.

Barnes, H. (1957). Process of restoration and synchronisation in marine ecology. The spring diatom increase and the "spawning" of the common barnacle *Balanus balanoides* L. *Année biol.* **33**, 67–85.

Barnes, H. and Powell, H. T. (1950). Some observations on the effect of fibrous glass surfaces upon the settlement of certain sedentary marine organisms. *J. mar. biol. Ass. U.K.* **29**, 299–302.

Barnes, H., Crisp, D. J. and Powell, H. T. (1951). Observations on the orientation of some species of barnacles. *J. Anim. Ecol.* **20**, 227–41.

Bayne, B. L. (1963). Responses of *Mytilus edulis* larvae to increase in hydrostatic pressure. *Nature, Lond.* **198**, 406.

Bayne, B. L. (1964a). The responses of the larvae of *Mytilus edulis* L. to light and to gravity. *Oikos* **15** (**1**), 162–74.

Bayne, B. L. (1964b). Primary and secondary settlement in *Mytilus edulis* L. (Mollusca). *J. Anim. Ecol.* **33**, 513–23.

Bayne, B. L. (1969). The gregarious behaviour of the larvae of *Ostrea edulis* L. at settlement, *J. mar. biol. Ass. U.K.* **49**, 327–56.

Bergquist, P. R. and Sinclair, M. E. (1968). The morphology and behaviour of larvae of some intertidal sponges. *N.Z. J. mar. freshwat. Res.* **2** (**3**), 426–37.

Bergquist, P. R., Sinclair, M. E. and Hogg, J. J. (1970). Adaptation of intertidal existence: reproductive cycles and larval behaviour in Demospongiae. *Symp. zool. Soc. Lond.* **25**, 247–71.

Berrill, N. J. (1947). The development and growth of *Ciona*. *J. mar. biol. Ass. U.K.* **26**, 616–25.

Berrill, N. J. (1950). "The Tunicata", 354 pp. Ray Society, London.

Bertholf, L. M. and Mast, S. O. (1944). Metamorphosis in the larvae of the tunicate *Styela partita*. *Biol. Bull. mar. biol. Lab. Woods Hole* **87**, 166.

Boaden, P. J. S. and Erwin, D. G. (1971). *Turbanella hyalina* versus *Protodriloides symbioticus*: a study in intertstitial ecology. *Vie Milieu, Supplement No.* **22**. Proc. Third European Marine Biology Symposium, 478–92.

Board, P. A. and Collings, T. M. (1965). Physical forces in marine fouling. *Discovery*, May 1965.

Bourdillon, A. (1954). Mise en évidence d'une substance favourisant la metamorphose des larves d'*Alcyonium corralloides* (Von Koch). *C.r. hebd. Seanc. Acad. Sci., Paris* **239**, 1434–6.

Bousfield, E. L. (1955). Ecological control of the occurrence of barnacles in the Miramichi Estuary. *Bull. natn. Mus. Can.* 137, 69 pp.

Bradway, W. (1936). The experimental alteration of the rate of metamorphosis in the tunicate *Clavelina huntsmani* (Van Name). *J. exp. Zool.* **72**, 213–225.

Burton, M. (1949). Observations on littoral sponges including the supposed swarming of larvae, movement and coalescence in mature individuals, longevity and death. *Proc. zool. Soc. Lond.* **118**, 893–915.

Carlisle, D. B. (1961). Locomotory powers of adult ascidians. *Proc. zool. Soc. Lond.* **136**, 141–146.

Carpine, C. (1960). Révision des connaissances sur la biologie et la prévention des Xylophages marins. *Bull. Inst. Océanogr. Monaco*, **1187**, 50 pp.

Carriker, M. R. (1951). Ecological observations of the distribution of oyster larvae in New Jersey estuaries. *Ecol. Monogr.* **21**, 19–38.

Carriker, M. R. (1956). Biology and propagation of young hard clams *Mercenaria mercenaria*. *J. Elisha Mitchell scient. Soc.* **72**, 57–60.

Cole, H. A. (1938). The fate of the larval organs in the metamorphosis of *Ostrea edulis*. *J. mar. biol. Ass. U.K.* **22**, 469–84.

Cole, H. A. and Knight-Jones, E. W. (1939). Some observations and experiments on the setting behaviour of *Ostrea edulis*. *J. Cons. perm. int. Explor. Mer* **14**, 86–105.

Cole, H. A. and Knight-Jones, E. W. (1949). The setting behaviour of larvae of the European oyster *Ostrea edulis* L. and its influence on methods of cultivation and spat collection. *Fishery Invest. Lond. Ser. II*, **17** (**3**), 39 pp.

Colman, J. S. (1933). The nature of the intertidal zonation of plants and animals. *J. mar. biol. Ass. U.K.* **18**, 435–76.

Colman, J. S. (1950). "The Sea and its Mysteries". Bell, London.

Cranfield, H. J. (1971). Studies on temporary attachment and cementing of *Ostrea edulis*. Ph.D. Thesis, University of Wales.

Crisp, D. J. (1953). Changes in the orientation of barnacles of certain species in relation to water currents. *J. Anim. Ecol.* **22**, 331–43.

Crisp, D. J. (1955). The behaviour of barnacle cyprids in relation to water movement over a surface. *J. exp. Biol.* **32**, 569–90.

Crisp, D. J. (1956). A substance promoting hatching and liberation in young in cirripedes. *Nature, Lond.* **178**, 263.

Crisp, D. J. (1958). The spread of *Elminius modestus* Darwin in North-West Europe. *J. mar. biol. Ass. U.K.* **37**, 483–520.

Crisp, D. J. (1961). Territorial behaviour in barnacle settlement. *J. exp. Biol.* **38**, 429–46.

Crisp, D. J. (1962). Swarming of planktonic organisms. *Nature, Lond.* **193**, 597–98.

Crisp, D. J. (1965). Surface chemistry, a factor in the settlement of marine invertebrate larvae. *Botanica Gothenburgensia* III. Proc. 5th Marine Biological Symposium, Goteborg, 1964, 51–65.

Crisp, D. J. (1967). Chemical factors inducing settlement in *Crassostrea virginica* Gmelin. *J. Anim. Ecol.* **36**, 329–35.

Crisp, D. J. and Austin, A. P. (1960). The action of copper in anti-fouling paints. *Ann. appl. Biol.* **48**, 787–99.

Crisp, D. J. and Barnes, H. (1954). The orientation and distribution of barnacles at settlement with particular reference to surface contour. *J. Anim. Ecol.* **23**, 142–62.

Crisp, D. J. and Ghobashy, A. F. A. A. (1971). Responses of the larvae of *Diplosoma* listerianum to light and gravity. *In* "Fourth European Marine Biology Symposium" (D. J. Crisp, ed.), pp. 443–465. Cambridge University Press, Cambridge.

Crisp, D. J. and Knight-Jones, E. W. (1953). The mechanism of aggregation in barnacle populations. *J. Anim. Ecol.* **22**, 360–62.

Crisp, D. J. and Meadows, P. S. (1962). The chemical basis of gregariousness in cirripedes. *Proc. R. Soc.* **B 156**, 500–20.

Crisp, D. J. and Meadows, P. S. (1963). Absorbed layers: the stimulus to settlement in barnacles. *Proc. R. Soc.* **B 158**, 364–87.

Crisp, D. J. and Patel, B. S. (1958). Relation between breeding and ecdysis in cirripedes. *Nature, Lond.* **181**, 1078–9.

Crisp, D. J. and Ritz, D. A. (1973). Responses of cirripede larvae to light. I. Experiments with white light. *Marine Biology* **23**, 327–335.

Crisp, D. J. and Ryland, J. S. (1960). Influence of filming and of surface texture on the settlement of marine organisms. *Nature, Lond.* **185**, 119.

Crisp, D. J. and Spencer, C. P. (1958). The control of the hatching process in barnacles. *Proc. R. Soc.* **B 148**, 278–99.

Crisp, D. J. and Stubbings, H. G. (1957). The orientation of barnacles to water currents. *J. Anim. Ecol.* **26**, 179–96.

Crisp, D. J. and Williams, G. B. (1960). Effects of extracts from fucoids in promoting settlement of epiphytic Polyzoa. *Nature, Lond.* **188**, 1206–7.

Cushing, D. H. (1963). Studies on a *Calanus* patch. V. The production cruises of 1954: summary and conclusions. *J. mar. biol. Ass. U.K.* **43**, 387–90.

Daniel, A. (1955). The primary film as a factor in settlement of marine foulers. *J. Madras Univ.* **25B**, 189–200.

Day, J. H. and Wilson, D. P. (1934). On the relation of the substratum to the metamorphosis of *Scolecolepis fuliginosa* (Claparède). *J. mar. biol. Ass. U.K.* **19**, 655–62.

Delage, Y. (1892). Embryonogenie des Eponges. *Arch. Zool. exp. gen.* **10**, 345–498.

De Silva, P. H. D. H. (1962). Experiments on the choice of substrate by *Spirorbis* larvae (Serpulidae). *J. exp. Biol.* **39**, 483–490.

Digby, P. S. B. (1961a). Mechanism of sensitivity of hydrostatic pressure in the prawn, *Palaemonetes varians* Leach. *Nature, Lond.* **191**, 366–68.

Digby, P. S. B. (1961b). Sensitivity of the shrimp *Crangon vulgaris* to hydrostatic pressure. *J. Physiol.* **158**, 12–13P.

Doochin, H. and Smith, F. G. W. (1951). Marine boring and fouling in relation to velocity of water currents. *Bull. mar. Sci. Gulf Caribb.* **1**, 196–208.

Duerden, J. E. (1899). The Edwardsia stage of the actinian *Lebrunia* and the formation of the gastro-colonic cavity. *J. Linn. Soc.* **27**, 269–316.

Duerden, J. E. (1905). The coral *Siderastria radians* and its post larval development. *Carnegie Inst. Wash. Pub.* **20**, 1–130.

Dybern, B. I. (1962/3). Biotope choice in *Ciona intestinalis* (L.). Influence of light. *Zool. Bidr. Upps.* **35**, 589–601.

Edmondson, C. H. (1929). Growth of Hawaiian corals. *Occ. Pap. Bernice P. Bishop Mus.* **58**, 1–38.

Edmondson, C. H. (1946). Behaviour of coral planulae under altered saline and thermal conditions. *Occ. Pap. Bernice P. Bishop Mus.* **18**, 283–304.

Evans, L. V. and Christie, A. O. (1970). Studies on the ship fouling alga *Enteromorpha*. I. Aspects of the fine structure and biochemistry of swimming and newly settled zoospores. *Ann. Bot.* **34**, 451–66.

Ewald, W. E. (1912). On artificial modifications of light reactions and the influence of electrolytes on phototaxis. *J. exp. Zool.* **13**, 591–612.

Floodgate, G. D. (1972). The mechanisms of bacterial attachments by detritus in aquatic systems. In press. *Memoria Instituto Italiano di Idrobiologie.*

Forbes, L., Seward, M. J. B. and Crisp, D. J. (1971). Orientation to light and the shading response in barnacles. *In* "Fourth European Marine Biology Symposium" (D. J. Crisp, ed.), pp. 539–58. Cambridge University Press, Cambridge.

Fraenkel, G. and Gunn, D. (1940). "The Orientation of Animals". Monographs on Animal Behaviour, 352 pp. Oxford.

Fretter, V. and Graham, A. (1962). "British Prosobranch Molluscs". Chapter 18, pp. 448–76. Ray Society, London.

Fry, W. G. (1971). The biology of larvae of *Ophlitaspongia seriata* (Grant, 1826) (Desmospongiae) from two North Wales populations. *In* "Fourth European Marine Biology Symposium" (D. J. Crisp, ed.), pp. 155–78. Cambridge University Press Cambridge.

Gabbott, P. A. and Larman, V. N. (1971). Electrophoretic examination of partially purified extracts of *Balanus balanoides* containing a settlement inducing factor. *In* "Fourth European Marine Biology Symposium" (D. J. Crisp, ed.), pp. 143–53. Cambridge University Press, Cambridge.

Gautier, Y. Y. (1954). Sur L'*Electra pilosa* des feuilles de Posidonies. *Vie Milieu* **5**, 65–70.

Gee, J. M. (1965). Chemical stimulation of settlement in larvae of *Spirorbis rupestris* (Serpulidae). *Anim. Behav.* **13**, 181–86.

Gee, J. M. and Knight-Jones, E. W. (1962). The morphology and larval behaviour of a new species of *Spirorbis* (Serpulidae). *J. mar. biol. Ass. U.K.* **42**, 641–54.

Ghobashy, A. F. A. A. (1967). Larval behaviour and metamorphosis in *Diplosoma listerianum*. Ph.D. Thesis, University of Wales.

Giard, A. (1876). Note sur l'embryogénie de la *Salmacine dysteri* Huxley. *C. r. hebd. Séanc. Acad. Sci. Paris* **82**, 233–35, 285–88.

Gibson, P. H. and Nott, J. A. (1971). Concerning the fourth antennular segment of the cypris larva of *Balanus balanoides*. *In* "Fourth European Marine Biology Symposium" (D. J. Crisp, ed.), pp. 227–36. Cambridge University Press, Cambridge.

Grave, B. H. (1930). The natural history of *Bulgula flabellata* at Woods Hole, Massachusetts, including the behaviour at attachment of the larva. *J. Morph.* **49**, 355–83.

Grave, C. (1920). *Amaroucium pellucidum* (Leidy) form *constellatum* (Verrill). 1. The activities and reactions of the tadpole larva. *J. exp. Zool.* **30**, 239–57.

Grave, C. (1921). *Amaroucium constellatum* (Verrill). II. The structure and organisation of the tadpole larva. *J. Morph.* **36**, 71–101.

Grave, C. (1926). *Molgula citrina* (Alder and Hancock). Activities and structure of the free swimming larva. *J. Morph.* **42**, 453–71.

Grave, C. (1932). The Botryllus type of ascidian larva. *Pap. Tortugas Lab.* **28** (**9**), 145–56.

Grave, C. (1935). Metamorphosis of ascidian larvae. *Pap. Tortugas Lab.* **29**, 209–92.

Grave, C. (1941). The "eye spots" and light responses of the larva of *Cynthia partita*. Abstract. *Biol. Bull. mar. biol. Lab. Woods Hole* **81**, 287.

Grave, C. (1944). The larva of *Styela* (*Cynthia*) *partita*. *J. Morph.* **75**, 173–91.

Grave, C. and McCash, G. K. (1923). *Perophora viridis*. The activities and structure of the free-swimming larva. *Wash. Univ. Stud. scient. Ser.* **11**, 89–116.

Gray, J. S. (1966a). Selection of sands by *Protodrilus symbioticus* (Giard). *Veroff. Inst. Meeresforsch. Bremerh.* **2**, 105–116.

Gray, J. S. (1966b). The attractive factor of intertidal sands to *Protodrilus symbioticus*. *J. mar. biol. Ass. U.K.* **46**, 627–45.

Gray, J. S. (1966c). Factors controlling the localizations of populations of *Protodrilus symbioticus* Giard. *J. Anim. Ecol.* **35**, 432–42.

Gray, J. S. (1967a). Substrate selection by the archiannelid *Protodrilus hypoleucus* Armenante. *J. exp. mar. Biol. Ecol.* **1,** 47–54.

Gray, J. S. (1967b). Substrate selection by the archianellid *Protodrilus rubropharyngeus* Jägersten. *Helgoländer wiss. Meeresunters.* **15**, 253–69.

Gray, J. S. (1968). An experimental approach to the ecology of the harpacticoid *Leptastacus constrictus* Lang. *J. exp. mar. Biol. Ecol.* **2**, 278–92.

Gray, J. S. (1971). Factors controlling population localizations in polychaete worms. *Vie Milieu.* Supplement No. 22. Proc. 3rd European Marine Biology Symposium, Arcachon, September, 1968. pp. 707–22.

Gray, J. S. and Johnson, R. M. (1970). The bacteria of a sandy beach as an ecological factor affecting the interstitial gastrotrich *Turbanella hyalina* Schultze. *J. exp. mar. Biol. Ecol.* **4**, 119–33.

Gregg, J. H. (1945). Background illumination as a factor in the attachment of barnacle cyprids. *Biol. Bull. mar. biol. Lab. Woods Hole* **88**, 44–49.

Hardy, A. C. and Bainbridge, R. (1951). Effect of pressure on the behaviour of decapod larvae. *Nature, Lond.* **167**, 354–55.

Harington, C. R. (1921). A note on the physiology of the ship-worm (*Teredo norvegica*). *Biochem. J.* **15**, 736–41.

Haskin, H. H. (1964). The distribution of oyster larvae. *Symp. exp. mar. Ecol. Occasional Pub.* **2**, 76–80. Naragansett Marine Laboratory.

Hawes, F. B. (1958). Preliminary observations on the settlement of the actinula larva of *Tubularia larynx* (Ellis and Scolender). *Ann. Mag. nat. Hist.* **13**, 147–55.

Hidu, J. (1969). Gregarious setting in the American oyster *Crassostrea virginica* Gmelin. *Chesapeake Sci.* **10** (**2**), 85–92.

Hilen, E. M. J. (1923). Report on a bacteriological study of ocean slime. *Bur. const. and Repair* (Navy Dept) (Quoted from Miller, Rapean and Whedon, 1948).

Hopkins, A. E. (1935). Attachment of larvae of the Olympia oyster, *Ostrea lurida*, to plane surfaces. *Ecology* **16**, 82–86.

Hopkins, A. E. (1937). Experimental observations on spawning, larval development and setting in the Olympia oyster, *Ostrea lurida*. *Bull. Bur. Fish., Wash.* **48**, No. 23.

Huxley, J. S. (1928). Experimentally induced metamorphosis in *Echinus* *Am. Nat.* **62**, 363–77.

Isham, L. B. and Tierney, J. Q. (1953). Some aspects of the larval development and metamorphosis of *Teredo pedicellata*. *Bull. mar. Sci. Gulf Caribb.* **2**, 574.

Isham, L. B., Walton Smith, F. G. and Springer, V. (1951). Marine borer attack in relation to conditions of illumination. *Bull. mar. Sci. Gulf Caribb.* **1**, 46–63.

Jägersten, G. (1940). Die Abhangigkeit der Metamorphose vom Substrat des Biotops bei *Protodrilus*. *Ark. Zool.* (*Stockholm*), **32A**, No. 17, 1–12.

James, B. L. (1971). Host selection and ecology of marine Digenean larvae. *In* "Fourth European Marine Biology Symposium" (D. J. Crisp, ed.), pp. 179–196. Cambridge University Press, Cambridge.

Kanda, S. (1919). On the reversibility of the heliotropism of *Arenicola* larvae by chemicals. *Biol. Bull. mar. biol. Lab. Woods Hole* **36**, 149–66.

Kato, M., Nakamura, K., Hirae, E. and Kakinuma, Y. (1961). The distribution pattern of Hydrozoa on seaweed with some notes on the so-called coaction among hydrozoan species. *Bull. Biol. Stn Asamushi* **10**, 195–202.

Kato, M., Hirae, E., and Kakinuma, Y. (1963). Further experiments on the interspecific relation in the colony formation among some hydrozoan species. *Sci. Rep. Tôhoku Univ. Ser.* 4 (*biol.*) **29**, 317–25.

Kato, M., Hirae, E. and Kakinuma, Y. (1967). Experiments on the coaction among hydrozoan species in the colony formation. *Sci. Rep. Tôhoku Univ. Ser.* 4 (*Biol.*) **33**, 359–73.

Kawagati, S. (1941). On the physiology of reef corals. V. Tropisms of coral planulae, considered as a factor of distribution on the reefs. *Palao trop. biol. Stn Stud.* **2**, 319–28.

Knight-Jones, E. W. (1951). Gregariousness and some other aspects of the setting behaviour of *Spirorbis*. *J. mar. biol. Ass. U.K.* **30**, 201–22.

Knight-Jones, E. W. (1952). Reproduction of oysters in the Rivers Crouch and Roach, Essex, during 1947, 1948 and 1949. *Fishery Invest. Lond., Ser.* 2 **18**, No. 2, 47 pp.

Knight-Jones, E. W. (1953a). Decreased discrimination during settling after prolonged planktonic life in larvae of *Spirorbis borealis* (Serpulidae). *J. mar. biol. Ass. U.K.* **32**, 337–45.

Knight-Jones, E. W. (1953b). Laboratory experiments on gregariousness during setting in *Balanus balanoides* and other barnacles. *J. exp. Biol.* **30**, 584–98.

Knight-Jones, E. W. (1955). The gregarious setting reaction of barnacles as a measure of systematic affinity. *Nature, Lond.* **174**, 266.

Knight-Jones, E. W. and Crisp, D. J. (1953). Gregariousness in barnacles in relation to the fouling of ships and to anti-fouling research. *Nature, Lond.* **171**, 1109.

Knight-Jones, E. W. and Morgan, E. (1966). Responses of marine animals to changes in hydrostatic pressure. *Oceanogr. mar. Biol. A. Rev.* **4**, 267–99.

Knight-Jones, E. W. and Moyse, J. (1961). Intraspecies competition in sedentary marine animals. *Symp. Soc. exp. Biol.* **15**, 72–95.

Knight-Jones, E. W. and Qasim, S. Z. (1955). Responses of some marine animals to changes in hydrostatic pressure. *Nature, Lond.* **175**, 941.

Knight-Jones, E. W. and Qasim, S. Z. (1966). Responses of Crustacea to changes in hydrostatic pressure. *In* "Proceedings of the Symposium on Crustacea, Marine Biological Association of India", pp. 1132–50.

Knight-Jones, E. W. and Stevenson, J. P. (1950). Gregariousness during settlement in the barnacle *Elminius modestus* Darwin. *J. mar. biol. Ass. U.K.* **29**, 281–97.

Knight-Jones, E. W., Bailey, J. H., and Isaac, M. S. (1971). Choice of algae by larvae of *Spirorbis*, especially *Spirorbis spirorbis*. *In* "Fourth European Marine Biology Symposium" (D. J. Crisp, ed.), pp. 89–104. Cambridge University Press, Cambridge.

Korringa, P. (1952). Recent advances in oyster biology. *Q. Rev. Biol.* **27**, 266–308, and 339–65.

Kunkle, D. E. (1957). The vertical distribution of oyster larvae in Delaware Bay. *Proc. natn. Shellfish. Ass.* **18**, 90–91.

Lance, J. (1962). Effects of water of reduced salinity on the vertical migration of zooplankton. *J. mar. biol. Ass. U.K.* **42**, 131–54.

Loeb, J. (1893). Ueber künstliche Umwandlung positiv heliotropischer Thiere in negativ heliotropische und umgekehrt. *Pflügers Arch. ges. Physiol.* **54**, 81–107.

Lucas, C. E. (1936). On certain interrelations between phytoplankton and zooplankton under experimental conditions. *J. Cons. perm. int. Explor. Mer* **11**, 343.

Lutaud, G. (1953). Premiers stedes de la croisieme zoariale chez le bryozoaire cheilostome *Scrupocellaria reptans* Thompson. *Archs Zool. exp. gén.* **90**, 42–58.

Lynch, W. F. (1947). The behaviour and metamorphosis of the larva of *Bulgula neritina* (Linnaeus): Experimental modification of the length of the free-swimming period and the responses of the larvae to light and gravity. *Biol. Bull. mar. biol. Lab. Woods Hole* **92**, 115–50.

Lynch, W. F. (1949a). Acceleration and retardation of the onset of metamorphosis in two species of *Bugula* from the Woods Hole region. *J. exp. Zool.* **3**, 27–55.

Lynch, W. F. (1949b). Modification of the responses of two species of *Bugula* larvae from Woods Hole to light and gravity. Ecological aspects of the behavior of *Bugula* larvae. *Biol. Bull. mar. biol. Lab. Woods Hole* **97**, 302–10.

Lynch, W. F. (1952). Factors influencing metamorphosis of *Bugula* larvae. *Biol. Bull. mar. biol. Lab. Woods Hole* **103**, 369–83.

Lynch, W. F. (1956). Experimental modification of the rate of metamorphosis of *Bugula* larvae. *J. exp. Zool.* **133**, 589–612.

Lynch, W. F. (1958). The effects of certain organic compounds and anti-mitotic agents on metamorphosis of *Bugula* and *Amaroecium* larva. *J. exp. Zool.* **137**, 117–52.

Lynch, W. F. (1960). Problems of the mechanism involved in the metamorphosis of *Bugula* and *Amaroucium* larvae. *Proc. Iowa Acad. Sci.* **67**, 522–31.

McArthur, R. H. and Wilson, F. O. (1967). "The Theory of Island Biogeography", 203 pp. Princeton Univ. Press, N.J.

McDougall, K. D. (1943). Sessile marine invertebrates of Beaufort, N. Carolina. *Ecol. Monogr.* **13**, 323–74.

Manning, J. H. and Whaley, H. H. (1954). Distribution of oyster larvae and spat in relation to some environmental factors in a tidal estuary. *Proc. natn. Shellfish. Ass.* **48**, 56–65.

Mast, S. O. (1921). Reactions to light in the larvae of the ascidians, *Amaroucium constellatum* and *Amaroucium pellucidum* with special reference to photic orientation. *J. exp. Zool.* **34**, 149–87.

Matthews, A. (1917). The development of *Alcyonium digitatum* with some notes on the early colony formation. *Q. Jl. microsc. Sci.* **62**, 43–94.

Mawatari, S. (1951). The natural history of a common fouling bryozoan *Bugula neritina*. *Misc. Rep. Res. Inst. nat. Resour. Tokyo* **19–21**, 47–54.

Mawatari, S. (1952). On *Watersipora cucullata* (Busk). Anatomical study. *Misc. Rep. Res. Inst. nat. Resour. Tokyo* **28**, 17–27.

Meadows, P. S. and Williams, G. B. (1963). Settlement of *Spirorbis borealis* larvae on surfaces bearing films of micro-organisms. *Nature, Lond.* **198**, 610–11.

Mesnil, F. and Caullery, M. (1917). Un nouveau type de dimorphisme evolutif chez une Annélide Polychète le *Spio martinensis*. *C.r. Séanc. Soc. Biol.* **165**, 646–8.

Miller, M. A., Rapean, J. C. and Whedon, W. F. (1948). The role of slime films in the attachment of fouling organisms. *Biol. Bull. mar. biol. Lab. Woods Hole* **94**, 143–57.

Miller, R. H. (1953). Liverpool Marine Biological Committee memoirs on typical British marine plants and animals. No. 35, Liverpool Univ. Press, 78 pp.

Moore, H. B. (1933). Change of orientation of a barnacle after metamorphosis. *Nature, Lond.* **132**, 969–70.

Moore, H. B. (1939). The colonization of a new rocky shore at Plymouth. *J. Anim. Ecol.* **8**, 29–38.

Moyse, J. (1971). Settlement and growth pattern of the parasitic barnacle *Pyrgoma anglicum*. *In* "Fourth European Marine Biology Symposium" (D. J. Crisp, ed.), pp. 125–41. Cambridge University Press, Cambridge.

Müller, D. G. (1964). Die Beteiligung eines Berührungsreiges beim Festsetzen von Algenschwärmen auf dem Substrat. *Z. Bot.* **52**, 193–8.

Nelson, J. (1912). Report of the Dept. of Biol. of the N.J. Agric. Coll. exp. Stn for the year ending June 30, 1911.

Nelson, T. C. (1928). Pelagic dissoconchs of the common mussel, *Mytilus edulis,* with observations on the behaviour of the larvae of allied genera. *Biol. Bull. mar. biol. Lab. Woods Hole* **55**, 180–92.

Nelson, T. C. (1954). Observations of the behaviour and distribution of oyster larvae. *Proc. natn. Shellfish. Ass.* **45**, 23–28.

Nelson, T. C. and Perkins, E. B. (1930). The reactions of oyster larvae to currents and to salinity gradients. *Anat. Rec.* **47**, 288.

Nott, J. A. (1969). Settlement of barnacle larvae: surface structure of the antennular attachment disc by scanning electron microscopy. *Mar. Biol.* **2**, 248–51.

Nott, J. A. and Foster, B. A. (1969). On the structure of the antennular attachment organ of the cypris larva of *Balanus balanoides* (L.). *Phil. Trans. R. Soc.* **B 256**, 115–34.

Pomerat, C. M. and Reiner, E. R. (1942). The influence of surface angle and of light on the attachment of barnacles and other sedentary organisms. *Biol. Bull. mar. biol. Lab. Woods Hole* **82**, 14–25.

Pomerat, C. M. and Weiss, C. M. (1946). The influence of texture and composition of surface on the attachment of barnacles and of other sedentary organisms. *Biol. Bull. mar. biol. Lab. Woods Hole* **91**, 57–65.

Prenant, M. and Teissier, G. (1924). Notes éthologiques sur la faune marine sessile des environs de Roscoff: Cirripèdes Bryozoaires Hydraires. *Trav. Stn biol. Roscoff* **2**, 3–49.

Pritchard, D. W. (1953). Distribution of oyster larvae in relation to hydrographic conditions. *Proc. Gulf Caribb. Fish. Inst.* **1952**, 123–32.

Prytherch, H. F. (1929). Investigations of the physical conditions controlling spawning of oysters and the occurrence, distribution and setting of the oyster larvae in Milford Harbour, Connecticut. *Bull. Bur. Fish., Wash.* **44**, 428–503.

Prytherch, H. F. (1934). The role of copper in the setting, metamorphosis and distribution of the American oyster, *Ostrea virginica*. *Ecol. Monogr.* **4**, 49–107.

Pyefinch, K. A., and Downing F. S. (1949). Notes on the general biology of *Tubularia larynx* Ellis and Solander. *J. mar. biol. Ass. U.K.* **28**, 21–43.

Ranade, M. R. (1957). Reversal of phototaxis in the larvae of *Polydora pulchra* Carazzi (Polychaeta, Spionidae). *Nature, Lond.* **179**, 151–2.

Rancurel, P. (1951). A propos de la larve de *Teredo pelicellata* Quatrefages. *Bull. Lab. marit. Dinard* **34**, 18–25.

Rasmont, R. (1970). Some new aspects of the physiology of freshwater sponges. *In* "The Biology of the Porifera" (W. G. Fry, ed.). *Symp. zool. Soc. Lond.* **25**, 415–22.

Rasmussen, E. (1951). Faunistic and biological notes on marine invertebrates. II. The eggs and larvae of some Danish marine gastropods. *Vidensk. Meddr. dansk naturh. Foren.* **113**, 201–49.

Rice, A. L. (1964). Observations on the effects of changes of hydrostatic pressure on the behaviour of some marine animals. *J. mar. biol. Ass. U.K.* **44**, 163–75.

Ross, D. M. (1967). Behavioural and ecological relationships between sea anemones and other invertebrates. *Oceanogr. mar. Biol. A. Rev.* **5**, 291–316.

Ross, D. M. and Sutton, L. (1963). A sea anemone, a hermit crab and a shell—an ecological triangle. *In* "Proceedings of the Sixteenth International Congress of Zoology, Washington, 1963", 62.

Ryland, J. S. (1959). Experiments on the selection of algal substrates by polyzoan larvae. *J. exp. Biol.* **36**, 613–31.

Ryland, J. S. (1960). Experiments on the influence of light on the behaviour of polyzoan larva. *J. exp. Biol.* **37**, 783–800.

Ryland, J. S. (1962a). The effect of temperature on the photic responses of polyzoan larvae. *Sarsia.* **6**, 41–48.

Ryland, J. S. (1962b). The association between Polyzoa and algal substrata. *J. Anim. Ecol.* **31**, 331–38.

Ryland, J. S. and Stebbing, A. R. D. (1971). Settlement and orientated growth in epiphytic and epizoic bryozoans. *In* "Fourth European Marine Biology Symposium" (D. J. Crisp, ed.), pp. 105–26. Cambridge University Press, Cambridge.

Scheltema, R. S. (1961). Metamorphosis of the veliger larvae of *Nassarius obsoletus* (Gastropoda) in response to bottom sediment. *Biol. Bull. mar. biol. Lab. Woods Hole* **120**, 92–109.

Scheltema, R. S. (1971). The dispersal of the larvae of shoal water benthic invertebrate species over long distances by ocean currents. *In* "Fourth European Marine Biology Symposium" (D. J. Crisp, ed.), pp. 7–28. Cambridge University Press, Cambridge.

Schijfsma, K. (1935). Observations on *Hydractinia echinata* (Flem.) and *Eupagurus bernhardus* L. *Archs Néerl. Sci.* **I**, 261–314.

Segrove, F. (1944). The development of the serpulid *Pomatoceros triqueta* L. *Q. Jl. microsc. Sci.* **82**, 467.

Sieburth, J. M. (1968). The influence of algal antibiosis on the ecology of marine micro-organisms. *Adv. in Microbiology of the Sea*, **1**, 63–89.

Sigerfoos, C. P. (1907). Natural history, organisation and late development of the Teridinae, or ship-worms. *Bull. Bur. Fish., Wash.* **27**, 191–231.

Singarajah, K. V., Moyse, J. and Knight-Jones, E. W. (1967). The effect of feeding on the phototactic behaviour of cirripede nauplii. *J. exp. mar. Biol. Ecol.* **1**, 144–53.

Skerman, T. M. (1958), Notes on settlement in *Lepas* barnacles. *N.Z. Jl Sci.* **I**, 385–90.

Smith, F. G. W. (1946). Effect of water currents upon the attachment and growth of barnacles. *Biol. Bull. mar. biol. Lab. Woods Hole* **90**, 51–70.

Smith, F. G. W. (1948). Surface illumination and barnacle attachment. *Biol. Bull. mar. biol. Lab. Woods Hole* **94**, 33–39.

Stebbing, A. R. D. (1971). The epizoic fauna of *Flustra foliacea* (Bryozoa). *J. mar. biol. Ass. U.K.* **51**, 283–300.

Swennen, C. (1961). Data on distribution, reproduction and ecology of the nudibranchiate molluscs occurring in the Netherlands. *Neth. J. Sea Res.* **I**, 191–240.

Tardy, J. (1962). Observations et des expériences sur la métamorphose et la croissance de *Capellinia exigue*. *C.r. hebd. Séanc. Acad. Sci., Paris* **254**, 2242–44.

Teitelbaum, M. (1966). Behaviour and settling mechanism of planulae of *Hydractinia echinata*. *Biol. Bull. mar. biol. Lab. Woods Hole* **131**, 410–114.

Thompson, T. E. (1958). The natural history, embryology, larval biology and post larval development of *Adalaria proxima* (Alder and Hancock) (Gastropoda Opisthobranchia). *Phil. Trans. R. Soc.* **B 242**, 1–58.

Thompson, T. E. (1962). Studies on the ontogeny of *Tritonia hombergi* Cuvier (Gastropoda Opisthobranchia). *Phil. Trans. R. Soc.* **B 245**, 171–218.

Thompson, T. E. (1964). Grazing and the life cycles of British nudibranchs. *In* "Grazing in Terrestrial and Marine Environments". *Brit. Ecol. Soc. Symp.* **No. 4**. (D. J. Crisp, ed.), pp. 275–97. Blackwell, Oxford.

Thorson, G. (1946). Reproduction and larval development of Danish marine bottom invertebrates. *Meddr. Kommn Havunders Kbh. Ser. Plankton,* **4**, 1–523.

Thorson, G. (1950). Reproductive and larval ecology of marine bottom invertebrates. *Biol. Rev.* **25**, 1–45.

Thorson, G. (1961). Length of pelagic larval life in marine bottom invertebrates as related to larval transport by ocean currents. *In* "Oceanography". *Am. Ass. Adv. Sci. Publ. No.* **67**, 455–74.

Thorson, G. (1964). Light as an ecological factor in the dispersal and settlement of larvae of marine bottom invertebrates. *Ophelia,* **1**, 167–208.

Trim, A. R. (1941). Studies in the chemistry of the insect cuticle. I. Some general observations on certain arthropod cuticles with special reference to the characterization of the proteins. *Biochem. J.* **35**, 1088–98.

Veitch, F. P. and Hidu, H. (1971). Gregarious setting in the American oyster *Crassostrea virginca* Gmelin. I. Properties of a partially purified setting factor. *Chesapeake Sci.* **12** (**3**), 173–78.

Visscher, J. P. (1928a). Nature and extent of fouling of ships' bottoms. *Bull. Bur. Fish., Wash.* **43** (**2**), 193–252.

Visscher, J. P. (1928b). Reactions of the cyprid larva of barnacles at the time of attachment. *Biol. Bull. mar. biol. Lab. Woods Hole* **54**, 327–35.

Vovelle, J. (1965). Le tube de *Sabellaria alveolata* (L.) annélide polychète *Hermellidae* et son ciment. Etude ecologique expérimentale, histologique et histochemique. *Archs. Zool. exp. gén.* **106**, 1–187.

Walley, L. J. (1969). Studies on the larval structure and metamorphosis of *Balanus balanoides. Phil. Trans. R. Soc.* **B** 256, 237–80.

Warburton, F. E. (1966). The behaviour of sponge larvae. *Ecology,* **47**, 672–74.

Weiss, C. M. (1947). The effect of illumination and stage of tide on the attachment of barnacle cyprids. *Biol. Bull. mar. biol. Lab. Woods Hole* **93**, 240–9.

Williams, G. B. (1964). The effect of extracts of *Fucus serratus* in promoting settlement of larvae of *Spirorbis borealis* (Polychaeta). *J. mar. biol. Ass. U.K.* **44**, 397–414.

Williams, G. B. (1965). Observations on the behaviour of planula larvae of *Clava squamata. J. mar. biol. Ass. U.K.* **45**, 257–73.

Wilson, D. P. (1932). On the mitraria larva of *Owenia fusiformis* Delle Chiaje. *Phil. Trans. R. Soc.* **B 221**, 231–334.

Wilson, D. P. (1937). The influence of the substratum on the metamorphosis of *Notomastus* larvae. *J. mar. biol. Ass. U.K.* **22**, 227–43.

Wilson, D. P. (1948). The relation of the substratum to the metamorphosis of *Ophelia* larvae. *J. mar. biol. Ass. U.K.* **27**, 723–60.

Wilson, D. P. (1952). The influence of the nature of the substratum on the metamorphosis of the larvae of marine animals, especially the larvae of *Ophelia bicornis* Savigny. *Annls. Inst. océanogr., Monaco.* **27**, 49–156.

Wilson, D. P. (1954). The attractive factor in the settlement of *Ophelia bicornis* Savigny. *J. mar. biol. Ass. U.K.* **33**, 361–80.

Wilson, D. P. (1955). The role of micro-organisms in the settlement of *Ophelia bicornis* Savigny. *J. mar. biol. Ass. U.K.* **34**, 531–43.

Wilson, D. P. (1968). The settlement behaviour of the larvae of *Sabellaria alveolata. J. mar. biol. Ass. U.K.* **48**, 387–435.

Wilson, D. P. (1970a). Additional observations on larval growth and settlement in *Sabellaria alveolata. J. mar. biol. Ass. U.K.* **50**, 1–31.

Wilson, D. P. (1970b). The larvae of *Sabellaria spinulosa* and their settlement behaviour. *J. mar. biol. Ass. U.K.* **50**, 33–52.

Wilson, H. V. (1907). On some phenomena of coalescence and regeneration in sponges. *J. exp. Zool.* **5**, 245–8.

Wisely, B. (1958a). The settling and some experimental reactions of a bryozoan larva, *Watersipora cucullata* (Busk). *Aust. J. mar. freshwat. Res.* **9** (**3**), 362–71.

Wisely, B. (1958b). The development and settling of a serpulid worm, *Hydroides norvegica* Gunnerus (Polychaeta). *Aust. J. mar. freshwat. Res.* **9**, 351–61.

Wisely, B. (1960). Observations on the settling behaviour of larvae of the tubeworm *Spirorbis borealis* Daudin (Polychaeta). *Aust. J. mar. freshwat. Res.* **11**, 55.

Wood, E. J. F. (1955). Effect of temperature and rate of flow on some marine fouling organisms. *Aust. J. Sci.* **18**, 34–37.

Wood, L. and Hargis, W. J. (1971). Transport of bivalve larvae in a tidal estuary. *In* "Fourth European Marine Biology Symposium" (D. J. Crisp, ed.), pp. 29–44. Cambridge University Press, Cambridge.

Yokota, T. (1936). Attachment of larvae of the Japanese oyster, *Ostrea gigas* (In Japanese). *Fish. exp. Stn. Miyagi Prefecture, Yosyoku Kwai Si*, **6** (**11**).

Yonge, C. M. (1962). On the primitive significance of the byssus in the Bivalvia and its effects in evolution. *J. mar. biol. Ass. U.K.* **42**, 113–125.

Zelenky, C. H. (1905). The rearing of serpulid larvae with notes on the behaviour of the young animals. *Biol. Bull. mar. biol. Lab. Woods Hole* **8**, 308–312.

Zhinkin, L. N. (1938). Stimulation of metamorphosis in ascidian larvae. *C. r. Acad. Sci. Moscow* (Doklady). *N.S.* **18** (**3**), 213–6.

ZoBell, C. E. (1939). The role of bacteria in the fouling of submerged surfaces. *Biol. Bull. mar. biol. Lab Woods Hole* **77**, 302.

ZoBell, C. E. and Allen, E. C. (1935). The significance of marine bacteria in the fouling of submerged surfaces. *J. Bact.* **29** No. 3, 239–51.

Author Index

Subject and Species Index

B

C

H

P

U

V

W

X

Y

Z